C PROGRAMMING

C PROGRAMMING

Shimpy Goyal
Charul Dewan
Meenakshi Yadav
Preety

NAMYA PRESS.com

Published by

Namya Press

India: 213, Vardaan House, 7/28 Ansari Road, Daryaganj, Delhi-110002
USA: 10685-B Hazelhurst Dr. # 31013 Houston, Texas USA - 77043
Email: namyapress@gmail.com
Website: https://namyapress.com

Edition	:	First Published in 2023 (Namya Press)
Title	:	C Programming
Editors	:	Shimpy Goyal, Charul Dewan, Meenakshi Yadav, Preety
ISBN	:	978-1-956861-28-0
Copyright	:	© Shimpy Goyal, Charul Dewan, Meenakshi Yadav, Preety 2023 All Rights Reserved

Preface

C programming is a widely used general-purpose programming language that has a huge impact on the development of modern software. This language is the foundation of many computer systems and applications that we use every day, including operating systems, software development tools, and video games.

This book serves as a comprehensive guide to C programming language, starting from the basics and covering all the essential topics, including arrays, functions, pointers, structures, and file handling. It also includes basic algorithms such as sorting, searching, and finding the square root of a number.

The first chapter introduces the basics of programming, including computer languages, creating and running programs, the preprocessor, compilation process, and the role of the linker. Chapter two introduces the C language, including its history, basic structure, tokens, keywords, identifiers, constants, strings, special symbols, variables, data types, and input/output statements.

Chapter three covers operators and expressions, including arithmetic, relational, and logical operators, assignment operators, increment and decrement operators, bitwise and conditional operators, and operator precedence and associativity. Chapter four discusses control structures, including decision statements, loop control statements, and jump statements.

Chapter five introduces arrays, including one-dimensional arrays, two-dimensional arrays, and multi-dimensional arrays. Chapter six covers functions, including user-defined and built-in functions, storage classes, parameter passing, call by value, passing arrays to functions, recursion, and more.

Chapter seven focuses on strings, including arrays of characters, variable-length character strings, inputting character strings, character library functions, and string handling functions. Chapter eight covers pointers, including pointer basics, pointer arithmetic, pointers to pointers, generic pointers, array of pointers, functions returning pointers, dynamic memory allocation, pointers to functions, and pointers and strings.

Chapter nine introduces structures and unions, including structure definition, initialization, and accessing of structures, nested structures, arrays of structures, structures and functions, self-referential structures, unions, typedef, and enumerations. Chapter ten covers file handling, including command-line arguments, file modes, basic file operations, read, write, and append.

Chapter eleven includes basic algorithms such as finding factorial, Fibonacci series, searching, basic sorting algorithms, finding the square root of a number, array order reversal, and reversal of a string. Finally, the appendix includes C99 extensions, including an introduction to computer systems, computing environments, and stdarg.h, and unistd.h.

Whether you are a beginner or an experienced programmer, this book provides a comprehensive guide to C programming language, including all the essential topics, practical examples, and exercises. We hope that this book will help you to develop a better understanding of the C programming language and its applications in the modern world of computing.

Contents

1

Introduction to Programming

The world of programming is rapidly expanding, and more and more people are becoming interested in learning how to code. Whether you are a complete beginner or have some experience with programming, this book chapter on Introduction to Programming is designed to help you get started on your journey.

In this chapter, we will cover the fundamentals of programming, including computer languages, creating and running programs, preprocessor, compilation process, role of linker, invocation and execution of a program, and more. We will also discuss the various programming paradigms and their advantages and disadvantages, so that you can choose the right one for your needs.

1.1 COMPUTER LANGUAGES

Computer languages, also known as programming languages, are formal languages designed to communicate instructions to a computer in order to perform a specific task or solve a problem. There are many different computer languages, each with its own syntax, semantics, and set of rules.

Computer languages can be divided into three main categories: low-level languages, high-level languages, and scripting languages. Low-level languages, such as assembly language and machine language, are closer to the hardware of the computer and are used to write system software such as operating systems and device drivers. High-level languages, such as C, Java, and Python, are more abstract and provide a more user-friendly way to write programs. They are used to write application software such as desktop applications, games, and web applications. Scripting languages, such as JavaScript and Ruby, are used to write scripts that automate tasks or provide interactivity on web pages.

Computer languages can also be classified based on their programming paradigm. The four main programming paradigms are procedural, object-oriented, functional, and logical. Procedural languages, such as C and Pascal, use a step-by-step approach to problem-solving. Object-oriented languages, such as Java and Python, use objects and classes to organize and structure programs. Functional languages, such as Haskell and Lisp, use mathematical functions as their primary building blocks. Logical languages, such as Prolog, use a declarative approach to problem-solving, where the program defines the problem and the computer finds a solution.

Each computer language has its own strengths and weaknesses, and choosing the right language for a particular task depends on several factors, such as the problem being solved, the hardware and software environment, and the programmer's experience and preferences. Some languages are easier to learn and use than others, while others are more efficient or more expressive.

In summary, computer languages are formal languages used to communicate instructions to a computer. They can be classified based on their level, programming paradigm, and strengths and weaknesses. Choosing the right language for a particular task depends on several factors and requires careful consideration.

1.2 CREATING AND RUNNING PROGRAMS

Creating and running programs is an essential aspect of software development. The process of creating and running programs involves several steps, including setting up the development environment, writing the code, compiling the code, and executing the program. All the steps in detail are mention below:

Setting up the Development Environment

The first step in creating a C program is to set up the development environment. This involves installing a C compiler on your computer. There are several free and open-source C compilers available, such as GCC, Clang, and Turbo C. You can choose any of these compilers based on your preference and operating system.

Once you have installed the C compiler, you can create a new C program using a text editor such as Notepad or TextEdit. Save the program with a .c extension to indicate that it is a C program.

Writing Your First C Program

Now that you have set up the development environment, it's time to write your first C program. The first program that every programmer writes is the "Hello, World!" program. This program simply displays the message "Hello, World!" on the screen.

Here is the code for the "Hello, World!" program:

```
#include <stdio.h>
 int main()
{
 printf("Hello, World!");
 return 0;
}
```

Let's break down the code:

- The first line of the program includes the standard input-output library, which contains the printf() function used to display text on the screen.
- The main() function is the entry point for the program. Every C program must have a main() function.
- The printf() function is used to display the message "Hello, World!" on the screen.
- The return statement at the end of the main() function indicates that the program has completed successfully.

Compiling and Running Your Program

Once you have written your C program, you need to compile it to create an executable file that can be run on your computer. To compile your program, open the command prompt or terminal and navigate to the directory where your program is saved.

To compile the "Hello, World!" program, type the following command:

```
gcc -o helloworld helloworld.c
```

This command uses the gcc compiler to create an executable file called "helloworld" from the source code file "helloworld.c".

To run the program, type the following command:

```
./helloworld
```

This command executes the "helloworld" program, which displays the message "Hello, World!" on the screen.

1.3 PREPROCESSOR

In C programming, the preprocessor is a program that performs various operations before the actual compilation process. The preprocessor takes a C program as input and performs text substitution and macro expansion. It also includes header files and performs conditional compilation.

The preprocessor is a part of the C compiler, which is responsible for executing the preprocessor directives written in the C program. These directives are written with a hash symbol (#) at the beginning of each line, and they are executed before the actual compilation process begins.

Functions of Preprocessor

There are various functions of preprocessor, which are explained below:

a. Text substitution: The preprocessor substitutes the text defined in macros with the actual text during the preprocessing stage. Macros are defined using the #define directive, which defines a constant value or expression that can be used throughout the program.

For example, the following line defines a macro for the value of pi:

```
#define PI 3.14159
```

After the macro is defined, you can use it in your code, like this:

float circumference = 2 * PI * radius;

The preprocessor will replace the text "PI" with "3.14159" during the preprocessing stage.

b. Macro expansion: The preprocessor also expands macros, which are created using the #define directive. Macro expansion is the process of replacing a macro with its value. It is performed during the preprocessing stage.

For example, consider the following macro:

```
#define SQUARE(x) (x*x)
```

You can use this macro in your code like this:

```
int result = SQUARE(5);
```

The preprocessor will replace SQUARE(5) with (5*5), resulting in the following code:

```
int result = (5*5);
```

c. Header file inclusion: The preprocessor includes header files, which contain function declarations, variable declarations, and other information needed for the program. The #include directive is used to include header files.

For example, consider the following line:

```
#include <stdio.h>
```

This line includes the header file "stdio.h", which contains declarations for standard input and output functions, such as printf() and scanf().

d. Conditional compilation: The preprocessor also performs conditional compilation, which is used to compile different parts of the program based on certain conditions. The #if, #ifdef, #ifndef, #elif, and #endif directives are used for conditional compilation.

For example, the following code snippet will compile the first part if the variable "DEBUG" is defined, and the second part if it is not defined:

```
#ifdef DEBUG printf("Debugging information\n"); #endif
#ifndef DEBUG printf("No debugging information\n"); #endif
```

The preprocessor is an essential part of the C programming language, which performs various tasks before the actual compilation process begins. It performs text substitution, macro expansion, includes header files, and performs conditional compilation. The preprocessor directives are written with a hash symbol (#) at the beginning of each line, and they are executed before the actual compilation process begins. Understanding the preprocessor is crucial for writing efficient and error-free C programs.

1.4 EXECUTION PROCESS

Execution is a crucial step in the development of C programs. It refers to the process of translating the source code written in a high-level programming language into a machine-readable form that can be executed by the computer. To discuss the execution process in C programming, including the various steps involved, the tools used, and the common errors that may arise.

- **Preprocessing:** The first step process is preprocessing. During this step, the preprocessor scans the source code and performs several operations, such as including header files, expanding macros, and removing comments. Header files are files that contain declarations of functions, variables, and constants that are needed by the program. Macros are a convenient way of defining short, reusable code snippets. They are expanded by the preprocessor, which replaces the macro name with its definition. Comments are ignored by the compiler and are used to document the code.
- **Compilation:** The next step is compilation, where the source code is translated into assembly language. Assembly language is a low-level programming language that is specific to a particular computer architecture. The compiler generates assembly code that is optimized for the target machine, taking into account the size and speed of the processor, the amount of memory available, and other factors.
- **Assembly:** The third step is assembly, where the assembly code generated by the compiler is translated into machine code. Machine code is a binary representation of the program that can be executed directly by the computer's processor. The assembler generates machine code by converting each instruction in the assembly code into the corresponding binary representation.
- **Linking:** The final step process is linking. During this step, the linker combines the object code generated by the assembler with any necessary libraries to create the final executable file. Libraries are collections of precompiled code that can be linked with a program to provide additional functionality. The linker resolves references to external symbols and ensures that all the necessary components are present.
- **Common Errors:** There are several common errors that can occur during the compilation process. One of the most common is a syntax error, which occurs when the code violates the rules of the programming language. Another common error is a semantic error, which occurs when the code is syntactically correct but does not behave as expected. Linker errors can occur if the program references a symbol that is not defined or if multiple definitions of a symbol exist.

1.5 ROLE OF LINKER

The process of creating a program involves multiple stages such as writing the code, compiling, and linking. In C programming, the linker plays a critical role in the final stage of creating an executable program. Here, we will discuss the role of the linker in C programming.

What is a Linker?

A linker is a program that takes the compiled code from one or more source files and combines them into a single executable program. When you compile a C program, it is translated into machine language and stored in an object file. The linker is responsible for linking these object files together to create an executable program.

The Linking Process: The linking process involves the following steps:

- **Object files:** The first step in the linking process is to gather all the object files that are created during the compilation process. An object file is a file that contains machine code and data, as well as information about the symbols used in the program.
- **Symbol resolution:** The linker then resolves all the symbols used in the program. A symbol is a name used to refer to a variable, function, or other entity in the program. The linker matches the symbols used in the program with the symbols defined in the object files.
- **Relocation:** The next step is to relocate the code and data. The linker adjusts the addresses of the code and data so that they match the final addresses in the executable program.
- **Linking:** Finally, the linker links all the object files together into a single executable program. It combines the code and data from the object files into a single program and produces an executable file that can be run on the target system.

Importance of the Linker

The linker plays a critical role in the final stage of creating an executable program. It ensures that all the object files are linked together correctly, resolves any symbols used in the program, and relocates the code and data to their final addresses. Without the linker, it would be impossible to create an executable program.

Another important role of the linker is in the creation of dynamic libraries. A dynamic library is a collection of object files that can be loaded and linked at run-time. The linker is responsible for creating the dynamic library and ensuring that it is linked correctly at run-time.

1.6 INVOCATION AND EXECUTION OF A PROGRAM

In C programming, the process of invoking and executing a program involves several steps, including writing the source code, compiling the code, and running the executable file. In this section, we will explain these steps in detail with code syntax and examples.

a. Writing the source code: The first step in writing a C program is to create a new file and write the source code. A C program consists of a set of functions, and every program must have a main function. The main function is the entry point of the program, and the program execution starts from here. Here is an example of a simple C program that prints "Hello, world!" to the console:

```
#include <stdio.h>
int main()
{
printf("Hello, world!");
 return 0;
}
```

In this program, we include the standard input/output library using the **#include** directive. The **main()** function is defined, and it calls the **printf()** function to print the message "Hello, world!" to the console. The program then returns 0 to indicate a successful execution.

b. Compiling the code The next step is to compile the source code into an executable file. We use a compiler, such as GCC, to do this. The syntax for compiling a C program using GCC is as follows:

```
gcc -o output_file input_file
```

In this command, **output_file** is the name of the file that will be created when the program is compiled, and **input_file** is the name of the C file that contains the source code. For example, if we saved the above program in a file called **hello.c**, we can compile it using the following command:

```
gcc -o hello hello.c
```

This command will create an executable file called **hello**.

c. Running the program The final step is to run the compiled program. We can do this by typing the name of the output file (without the '.c' extension) in the command line. For example, if we compiled our program as **hello**, we can run it using the following command:

```
./hello
```

If everything is correct, we should see the output of our program on the screen, which should be "Hello, world!".

In summary, the process of invoking and executing a C program involves writing the source code, compiling the code using a compiler such as GCC, and running the resulting executable file in the command line. By following these steps, we can easily compile and execute our C programs, and begin exploring the many possibilities of this powerful programming language.

EXERCISE

Short Questions (2 marks each):

1. What are computer languages?
2. How do you create and run a program?
3. What is the role of a preprocessor in programming?
4. What is the compilation process in programming?
5. What is the role of a linker in programming?

Long Questions (5 marks each):

1. Explain the difference between interpreted and compiled programming languages.
2. Describe the steps involved in the compilation process of a program.
3. What is a preprocessor and how does it work? Give an example of a commonly used preprocessor in programming.
4. Explain the role of a linker in creating an executable program.
5. Discuss the process of program invocation and execution in detail.

Multiple Choice Questions

1. Which of the following is not a type of computer language?
   a) C++
   b) HTML
   c) CSS
   d) Hammer

Answer: d) Hammer

2. Which type of programming language is converted into machine code and executed directly by the computer?
   a) Interpreted
   b) Compiled
   c) Hybrid
   d) None of the above

Answer: b) Compiled

3. What is the purpose of a preprocessor in programming?
   a) To convert source code into machine code
   b) To identify errors in the code
   c) To expand macros and conditional compilation statements
   d) To execute the program

Answer:c) To expand macros and conditional compilation statements

4. What is the first step in the compilation process of a program?
   a) Linking
   b) Optimization
   c) Preprocessing
   d) Parsing

Answer: c) Preprocessing

5. Which of the following is not a role of a linker in programming?
   a) Resolving external references
   b) Generating object code
   c) Combining object files
   d) Allocating memory

Answer: b) Generating object code

6. Which of the following is an example of a commonly used preprocessor in programming?
    a) GNU Compiler Collection (GCC)
    b) Visual Studio Code
    c) Make
    d) C preprocessor (cpp)

Answer: d) C preprocessor (cpp)

7. What is the process of converting source code into machine code called?
    a) Linking
    b) Compiling
    c) Executing
    d) Debugging

Answer: b) Compiling

8. Which step in the compilation process generates an object file?
    a) Parsing
    b) Preprocessing
    c) Code generation
    d) Optimization

Answer: c) Code generation

9. What is the final step in the creation of an executable program?
    a) Compiling
    b) Linking
    c) Preprocessing
    d) Debugging

Answer: b) Linking

10. What is the process of loading a program into memory and executing it called?
    a) Invocation
    b) Compilation
    c) Preprocessing
    d) Linking

Answer: a) Invocation

2

Introduction to C Language

C is programming language developed in 1972 by Dennis M. Ritchie at the Bell Telephone Laboratories. It is reliable, capable & friendly computer programming language. C & Java programming language both are equally popular wherein C is used for Unix Programming and Java is mainly used among modern software programmers.

2.1 HISTORY OF C LANGUAGE

1960:	ALGOL-60, found at Cambridge University and developed for Tranjae Alternate language
1963:	CPL - Combined Programming Language
1967:	BCPL - Basic Combined Programming Language. Found by Martin Richards at Cambridge University.
1969:	B <- B Language - founded by Ken Thompson at Bell Labs
1972:	C language by Dennis Ritchie. He added here Datatype concept.
1989:	ANSI - Develop standard for C, also referred as “C89”
1990:	Approved by ISO, also called “C90”
1995:	ISO released extension for C for interse i.e. “C95”

1999: Revised by ISO i.e. named as "C99" Dec'2011 - "C11" standard.

Why use C?

At early stage C was basically used in development work of the system like in programs which build up the operating Systems. But later on it become a system development language because of its user friendly feature as it can be easily readable & writable in compare to assembly language as assembly language is difficult for human being to read & write. Some examples of the use of C might be –

- Language Compilers
- Assemblers
- Text Editors
- Network Drivers
- Modern Programs
- Databases
- Language Interpreters
- Operating Systems
- Utilities

The C Character set

Character set: The character set is the fundamental of any language and they are used to represent information. Like natural languages,computer language will also have well defined character set, which is useful to build the programs.

Character:- It denotes any alphabet, digit or special symbol used to represent information.These characters can be combined to form variables. C *uses constants, variables, operators, keywords and expressions as building blocks to form a basic C program.* The characters in C are grouped into the following two categories:

1. Source character set

- Alphabets
- Characters
- Digits Special

- White Spaces

2. Execution character set

- Escape Sequence

Source Character Set

ALPHABETS

Uppercase letters A-Z

Lowercase letters a-z

DIGITS

0, 1, 2, 3, 4, 5, 6, 7, 8, 9

SPECIAL CHARACTERS

<table>
<tr><td>~</td><td>tilde</td></tr>
<tr><td>%</td><td>percent sign</td></tr>
<tr><td>|</td><td>vertical bar</td></tr>
<tr><td>@</td><td>at symbol</td></tr>
<tr><td>+</td><td>plus sign</td></tr>
<tr><td><</td><td>less than</td></tr>
<tr><td>_</td><td>underscore</td></tr>
<tr><td>-</td><td>minus sign</td></tr>
<tr><td>></td><td>greater than</td></tr>
<tr><td>^</td><td>caret</td></tr>
<tr><td>#</td><td>number sign</td></tr>
<tr><td>=</td><td>equal to</td></tr>
<tr><td>&</td><td>ampersand</td></tr>
<tr><td>$</td><td>dollar sign</td></tr>
<tr><td>/</td><td>slash</td></tr>
<tr><td>(</td><td>left parenthesis</td></tr>
<tr><td>*</td><td>asterisk</td></tr>
<tr><td>\</td><td>back slash</td></tr>
<tr><td>)</td><td>right parenthesis</td></tr>
<tr><td>'</td><td>apostrophe</td></tr>
<tr><td>:</td><td>colon</td></tr>
</table>

[	left bracket
“	quotation mark
;	semicolon
]	right bracket
!	exclamation mark
,	comma
{	left flower brace
?	Question mark
.	dot operator
}	right flower brace

WHITESPACE CHARACTERS

\b	blank space
\t	horizontal tab
\v	vertical tab
\r	carriage return
\f	form feed
\n	new line
\\	Back slash
\’	Single quote
\”	Double quote
\?	Question mark
\0	Null
\a	Alarm (bell)

Execution Character Set

Certain ASCII characters are unprintable, means they are not displayed on the screen. Those characters perform other functions aside from displaying text. Examples are backspacing, moving to a newline, or ringing a bell.

These are used in output statements. Escape sequence usually consists of a backslash and a letter or a combination of digits. An escape sequence is considered as a single character but a valid character constant.

These are employed at the time of execution of the program.

Execution characters set are always represented by a backslash (\) followed by a character. Note that each one of character constants represents one character, although they consist of two characters. These characters combinations are called as **escape sequence**.

Backslash Character Constants

Character	ASCII Value	Escape Sequence	Result
Null	000	\0	Null
Alarm (bell)	007	\a	Beep Sound
Back space	008	\b	Moves previous position
Horizontal tab	009	\t	Moves next horizontal tab
New line	010	\n	Moves next Line
Vertical tab	011	\v	Moves next vertical tab
Form feed	012	\f	Moves initial position of next page
Carriage return	013	\r	Moves beginning of the line
Double quote	034	\”	Present Double quotes
Single quote	039	\’	Present Apostrophe
Question mark	063	\?	Present Question Mark
Back slash	092	\\	Present back slash
Octal number	\000		
Hexadecimal number	\x		

2.2 C IDENTIFIERS

Identifier refers to entities such as variables, functions, structures etc.

Identifier must be unique, it means unique name is being given to an entity so that it can be easily identified during the execution of the program. For example:

```
int age;
double roll _number;
```

Here *age* and roll _numbe rare identifiers.

Always remember, identifier names must be different from keywords. You cannot use *int* as an identifier because *int* is a keyword.

Rules of Creating an Identifier

- A valid identifier can have letters (both uppercase and lowercase letters), digits and underscores.
- The first letter of an identifier should be either a letter or an underscore.
- There is no rule on length of an identifier. However, the first 31 characters of identifiers are discriminated by the compiler.

2.3 KEYWORDS

Keywords are predefined, reserved words used in C programming that have special meanings to the compiler. These cannot be used as an identifier For example:

```
int roll_no;
```

Here, *int* is a keyword and *'roll_no'* is a variable of type integer.

As C is a case sensitive language, all keywords must be written in lowercase. There are 32 keywords available in C language Here is a list of all keywords allowed in ANSI C.

Keywords in C Language

int	auto	double	struct
break	else	long	switch
case	enum	typedef	register
char	extern	union	return
continue	for	signed	void
do	if	static	while
default	goto	sizeof	volatile
const	float	short	unsigned

2.4 CONSTANTS

A constant is a value or an identifier whose value cannot be altered in a program. For example: 2, 1.5, "Student_Name", etc.

As mentioned, an identifier also can be defined as a constant.

```
const double PI = 3.14
```

Here, *PI* is a constant. Basically what it means is that, *PI* and *3.14* is same for this program.

Below are the different types of constants you can use in C.

1. Integer constants

An integer constant is a numeric constant (associated with number) without any fractional or exponential part. It can be either positive or negative . The Range of integer constant is -32768 to +32767.There are three types of integer constants in C programming:

```
decimal constant(base 10)
octal constant(base 8)
hexadecimal constant(base 16)
```

For example:

```
Decimal constants: 0, -9, 22 etc
Octal constants: 021, 077, 033 etc
Hexadecimal constants: 0x7f, 0x2a, 0x521 etc
```

In C programming, octal constant starts with a 0 and hexadecimal constant starts with a 0x.

2. Floating-point constants

A floating point constant is a numeric constant that has either a fractional form or an exponent form. Range of floating point constant is expressed in exponential form is -3.4e38 to 3.4e38 Example:

```
-2.0
0.0000234
-0.22E-5
```

Note: E-5 = 10^{-5}

3. Character constants

A character constant is a constant which uses single quotation around characters.The maximum length for character consant is 1 character. For example: 'a', '1', 'm', 'F'

4. Escape Sequences

Sometimes, it is necessary to use characters which cannot be typed or has special meaning in C programming. For example: newline(enter), tab, question mark etc. In order to use these characters, escape sequence is used.

For example: `\n` is used for newline. The backslash (`\`) causes "escape" from the normal way the characters are interpreted by the compiler.

Escape Sequences	
Escape Sequences	**Character**
\b	Backspace
\f	Form feed
\n	Newline
\r	Return
\t	Horizontal tab
\v	Vertical tab
\\	Backslash
\'	Single quotation mark
\"	Double quotation mark
\?	Question mark
\0	Null character

5. String constants

String constants are the constants which are enclosed in a pair of double-quote marks. For example:

```
"good" //string constant
"" //null string constant
"      " //string constant of six white space
"x" //string constant having single character.
"Earth is round\n" //prints string with newline
```

6. Enumeration constants

Keyword enum is used to define enumeration types. For example:

```
enumcolor {yellow, green, black, white};
```

Here, *color* is a variable and *yellow, green, black* and *white* are the enumeration constants having value 0, 1, 2 and 3 respectively

2.5 VARIABLES

In programming, a variable is a container (storage area) to hold data. The content of variable can change . In C programming, you have to declare a variable before you use it.

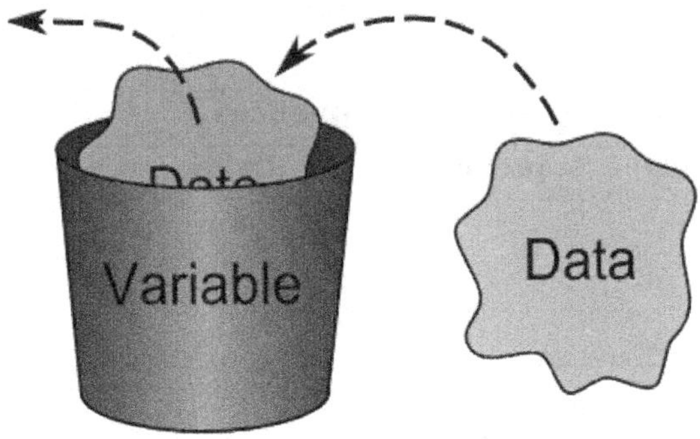

To indicate the storage area, each variable should be given a unique name. Variable names are just the symbolic representation of a memory location. For example:

```
int playerScore = 95;
```

Here, *playerScore* is a variable of integer type. The variable is assigned value: 95.

Rules for Naming a Variable in C

A variable name is any combination of 1 to 8 alphabets, (both uppercase and lowercase letters), digits and underscores. Some compilers allow variable names whose length could be upto 40 characters.but it would be safer to stick to the rule of 8 characters.

The first letter of a variable name must be an alphabet.

No commas or blanks are allowed within a variable name.

No special symbol other than an underscore(as in gross_sal) can be used in a variable name.

Example:

```
int si, m_hra;
float rate;
char code;
```

2.6 DATA TYPES

In C programming, variables or memory locations should be declared before it can be used. Similarly, a function also needs to be declared before use.

Data types simply refers to the type and size of data associated with variables and functions.

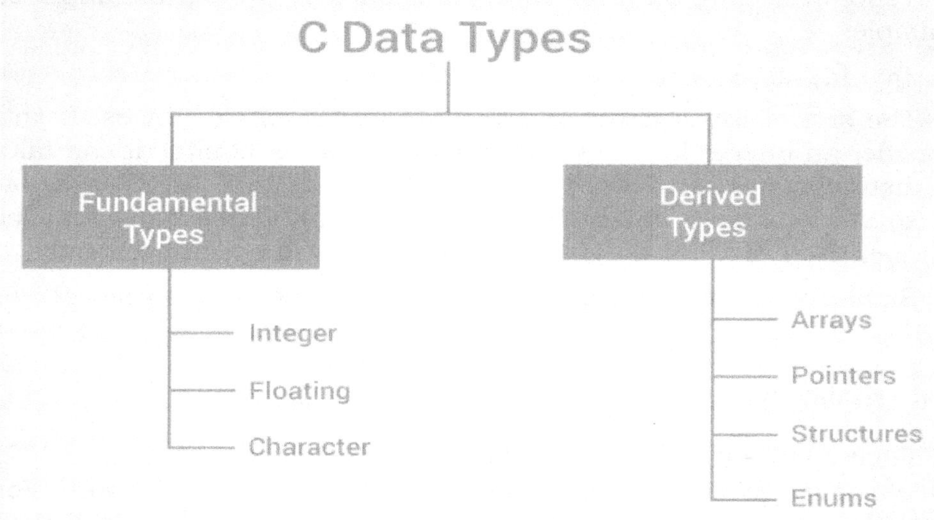

Figure: Flow Chart for Categorization of Data Types

Data Types in C

Fundamental Data Types :- are predefined datatypes.

- Integer types
- Floating type
- Character type

Derived Data Types :-These are defined by users for there convince.

- Arrays
- Pointers
- Structures
- Enumeration

First, we will discuss about fundamental data type & then we will discuss about derived data type.

```
int - Integer data types
```

Integers are whole numbers that can have both positive and negative values but no decimal values. Example: 0, -5, 10

In C programming, keyword *int* is used for declaring integer variable. For example:

```
int id;
```

Here, *id* is a variable of type integer.

You can declare multiple variable at once in C programming. For example:

```
int id, age;
```

The size of int is either 2 bytes(In older PC's) or 4 bytes. If you consider an integer having size of 4 byte(equal to 32 bits), it can take 2^{32} distinct states as: -2^{31},$-2^{31}+1$, ...,-2, -1, 0, 1, 2, ..., $2^{31}-2$, $2^{31}-1$. If you try to store larger number than $2^{31}-1$, i.e,+2147483647 and smaller number than -2^{31}, i.e, -2147483648, program will not run correctly.

Similarly, int of 2 bytes, it can take 2^{16} distinct states from -2^{15} to $2^{15}-1$.

float - Floating Types

Floating type variables can hold real numbers such as: 2.34, -9.382, 5.0 etc. It is declared in C by using either *float* or *double* keyword. For example:

```
Float accountBalance;
Double bookPrice;
```

Here, both *accountBalance* and *bookPrice* are floating type variables.

In C, floating values can be represented in exponential form as well. For example:

```
Float normalization Factor = 22.442e2;
```

Difference between Float and Double

The size of *float* (single precision float data type) is 4 bytes. And the size of *double* (double precision float data type) is 8 bytes. Floating point variables has a precision of 6 digits whereas the precision of double is 14 digits.

char - Character types

Keyword *char* is used for declaring character type variables. For example:

```
char test = 'h';
```

Here, *test* is a character variable. The value of *test* is 'h'.

The size of character variable is 1 byte.

Qualifiers

Qualifiers alters the meaning of base data types to yield a new data type.

Size qualifiers

Size qualifiers alters the size of a basic type. There are two size qualifiers, *long* and *short*. For example:

```
long double i;
```

The size of *double* is 8 bytes. However, when *long* keyword is used, that variable becomes 10 bytes.

There is another keyword *short* which can be used if you previously know the value of a variable will always be a small number.

Sign qualifiers

Integers and floating point variables can hold both negative and positive values. However, if a variable needs to hold positive value only, *unsigned* data types are used. For example:

```
// unsigned variables cannot hold negative value
Unsigned int positiveInteger;
```

There is another qualifier *signed* which can hold both negative and positive only. However, it is not necessary to define variable *signed* since a variable is signed by default.

An integer variable of 4 bytes can hold data from -2^{31} to $2^{31}-1$. However, if the variable is defined as unsigned, it can hold data from 0 to $2^{32}-1$.

It is important to note that, sign qualifiers can be applied to *int* and *char* types only.

Constant qualifiers

An identifier can be declared as a constant. To do so *const* keyword is used.

```
Const int cost = 20;
```

The value of *cost* cannot be changed in the program.

Volatile qualifiers

A variable should be declared volatile whenever its value can be changed by some external sources outside the program. Keyword *volatile* is used for creating volatile variables.

EXERCISE

Short Questions (2 marks)

1. What is the history of C language?
2. What is the basic structure of C programs?
3. What are C tokens?
4. What is the process of compiling and running a C program?
5. What are keywords in C language?

Long Questions (5 marks)

1. Explain the history of C language and its evolution.
2. Discuss the basic structure of C programs with the help of an example.
3. What are the different types of constants in C language? Explain with examples.
4. Discuss the process of compiling and running a C program.
5. Explain the different data types available in C language.

Multiple Choice Questions

1. Which of the following is a keyword in C language?
   a) function
   b) string
   c) printf
   d) none of the above

Answer: a) function

2. Which of the following is not a valid identifier in C language?
   a) 123abc
   b) abc123
   c) _abc123
   d) None of the above

Answer: a) 123abc

3. Which of the following is a special symbol in C language?
   a) +
   b) &
   c) $
   d) *

Answer: c) $

4. Which of the following is a data type in C language?
   a) if
   b) char
   c) break
   d) void

Answer: b) char

5. Which of the following is a constant in C language?
   a) scanf
   b) 5.5
   c) float
   d) double

Answer: b) 5.5

6. What is the syntax for declaring a variable in C language?
   a) var_name variable_type;
   b) variable_type var_name;
   c) var_name = variable_type;
   d) None of the above

Answer: b) variable_type var_name;

7. Which of the following is not a valid input/output statement in C language?
   a) scanf
   b) printf
   c) read
   d) None of the above

Answer: c) read

8. Which of the following is an example of a string in C language?
   a) 'a'
   b) 5
   c) "hello"
   d) None of the above

Answer: c) "hello"

9. What is the process of converting a float variable to an integer variable called in C language?
   a) Casting
   b) Declaration
   c) Initialization
   d) None of the above

Answer: a) Casting

10. Which of the following is not a step in the process of compiling and running a C program?
    a) Editing the source code
    b) Saving the source code
    c) Typing the program name in the compiler
    d) None of the above

Answer: d) None of the above (all the given options are steps in the process of compiling and running a C program)

3

Operators and Expressions

3.1 INTRODUCTION

C programming language is a widely-used and versatile language, which is popular among software developers, system programmers, and embedded systems engineers. The language has a rich set of features and constructs that allow programmers to create complex and efficient programs. One of the key aspects of C programming is the use of expressions and operators.

In this chapter, we will explore the various types of expressions and operators in C programming. We will discuss how to use these constructs to perform mathematical operations, compare values, and make decisions in a program. We will also explore some of the more advanced operators, such as bitwise operators, that allow for more efficient manipulation of data.

By the end of this chapter, you will have a solid understanding of expressions and operators in C programming. You will be able to use these constructs to write efficient and powerful programs, and you will have a foundation for learning more advanced topics in the language. Top of ForBottom of Form

3.2 EXPRESSIONS

Expressions is a sequence of operators and operands that specifies computation of a value. An expressions may consist of single entity or some combination of such entities interconnected by one or more operators.

All expression represents a logical connection that's either true or false.

Thus logical type expression actually represents numerical quantities.

In C every expression evaluates to a value i.e., every expression results in some value of a certain type that can be assigned to a variable. Some examples of expression are shown in the table given below.

```
A+b
3.14*r*r
a*a+2*a*b+b*b
```

In Programming, different verities of expressions are given to the compiler. Expressions can be classified on the basis of Position of Operators in an expression –

Type	Explanation	Example
Infix	Expression in which Operator is in between Operands	a + b
Prefix	Expression in which Operator is written before Operands	+ a b
Postfix	Expression in which Operator is written after Operands	a b +

These expressions are solved using the stack.

3.3 C – EXPRESSION EXAMPLE PROGRAM

```
#include<stdio.h>
#include<conio.h>
int main()
{
int a=123 , b;
b=a+2; // here a+2 form a expression.because a,2 are operand
& + is operator after evaluation values goes store in b .
clrscr();
printf("Value of a = %d\n",a);
printf("Value of b =%d",b);
    getch();
```

```
return 0;
}
```

Output:

```
Value of a = 123
Value of b = 125
```

3.4 OPERATORS

An operators is a symbol that helps us to perform specific mathematical and logical computations on operands..

For example, consider the below statement:

```
c = a + b;
```

Here, '+' is the operator known as *addition operator* and 'a' and 'b' are operands. The addition operator tells the compiler to add both of the operands 'a' and 'b'. C has many built-in operator types and they can be classified as:-

1. Arithmetic operators
2. Relational operators
3. Logical operators
4. Assignment operators
5. Conditional operators
6. Comma operators
7. Unary operators
8. Bitwise Operators

1. Arithmetic Operators

The Arithmetic operators perform Arithmetic operations . A list of arithmetic operator and their meaning are given below.

Operator Meaning

+ Addition or Unary plus

– Subtraction or Unary minus

* Multiplication

/ Division

% Modulo Division

This table shows the symbols of arithmetic, together with their duties. These operators allow you to write expressions whose evaluation is precisely the treatment of information that made the computer.

For example

Suppose that a and b are integer variables whose values are 100 and 4, respectively. Arithmetic expressions involving these variables are shown below, together with their resulting values.

Variable1	operator	variable2	Value
a	+	b	104
a	–	b	96
a	*	b	400
a	/	b	25
a	%	b	0

2. Relational Operators

Relational operators are used to compare, logical, arithmetic and character expression. Each of these six relational operators takes two operands. Each of these operators compares their left side with their right side. It evaluates to 0 if the condition is false and 1 if it is true.

Operator	Meaning
<	Less than
>	Greater than
<=	Less than or equal to
>=	Greater than or equal to
==	Equal to
!=	Not equal to

The concept of comparison certainly implies different meanings depending on the data type of the variables compared. Operators presented in this table are suitable for comparing numeric values but not to compare strings or lists of any kind. String comparison requires the use of features designed specifically for this purpose, like strcmp() these kind of operators are discussed in upcoming chapter .

Example of Relational Operator

Suppose that a and b are integer variables whose values are 100 and 4, respectively. Several relational operator involving these variables are shown below, together with their resulting values.

Expression	Interpretation	Value
a<b	False	0
a>b	True	1
a<=b	False	0
a>=b	True	1
a==b	False	0
a!=b	True	1

3. Logical Operators

A Logical operator is used to compare or evaluate logical and relational expression that the only values that can take operands are True or False.

There are three logical operators in C language. They are

Operator	Meaning
&&	AND
||	OR
!	NOT

Two versions of these operators are :-

- When you simply want to know if an expression **is true or false (eg, x>7),** operators are used **lógicos** . These **operators do not consider the structure of bits, simply take the value 0 as false and any other as true.**
- There is also another **version (the bitwise) imposed** by the logical operations to the bits used to internally represent the operands. These are **the operators & , | and ~ , respectively**.

Example of Logical Operator

Suppose that a and b are integer variables whose values are 100 and 4, respectively. Several logical operator with relational operator involving these variables are shown below, together with their resulting values.

Expression	Interpretation	Value
(a>b)&&(a==100)	True	1
(b>a)||(a>b)	True	1
!(a>b)	False	0

4. Assignment Operator

An Assignment operator is used to form an assignment expression, which assigns the value to an identifier. The most commonly used assignment operator is = . Assignment expressions that make use of this operator are written in the form

```
Identifier = expression;
```

Example:

```
i=100;
j=200;
```

Other forms of assignment operators exist that are obtaining various operators such as +,-,*,/,% etc with the = sign.

operator	Equation	Output
+=	i=i+1	i+=1
-=	i=i-10	i-=10
*=	i=i*11	i*=11
/=	i=i/12	i/=12
%=	i=i%5	i%=5

C – ASSIGNMENT OPERATORS EXAMPLE PROGRAM

```
#include<stdio.h>
#include<conio.h>
intmain()
{
int a=123,b; // here 123 is assigned to variable a .
b=a+2; // here a+2 after evaluatation is assigned to b.
clrscr();
printf("Value ofa =%d\n",a);
printf("Value ofb=%d",b);
getch( );
return0;
}
```

Output:

```
Value of a = 123
Value of b = 1
```

5. Conditional Operator

The conditional expression can be used as **shorthand for some if-else statements**.

Conditional operator is also **called as ternary operator.**

This **operator consist of two symbols: the question mark (?) and the colon (:).**

Syntax:

```
Identifier = (test expression)? Expression1: Expression2 ;
```

This is an expression, not a statement, so it represents a value. The operator works by evaluating test expression. If it is true (non-zero), it evaluates and returns expression1. Otherwise, it evaluates and returns expression2.

The classic example of the ternary operator is to return the smaller of two variables. Every once in a while, the following form is just what you needed. Instead of..

```
if(x<y)
{
min=x;
}
else
{
min=y;
}
```

Simply use

```
min = (x < y) ?x : y;
```

Suppose that x and y are integer variables whose values are 100 and 4, respectively. After executing above statement, the value of min is 4.

This is the only operator in C that makes use of three operands. It test condition corresponding to test expression, if it is true the value corresponding to Expression1 and it is false it corresponds to the value of Expression2.

C Conditional Operator Example Program

```
#include<stdio.h>
#include<conio.h>
int main()
{
int a,b,result,choice;
clrscr();
printf("Enter first number\n");
scanf("%d",&a);
printf("Enter second number\n");
scanf("%d",&b);
printf("Enter1foraddition or2formultiplication\n");
scanf("%d",&choice);
result=(choice==1)?a+b:((choice==2)?a*b:printf("Invalid
Input"));// Conditional operator evlautation
if(choice==1||choice==2)
printf("The result is%d\n\n",result);
getch();
}
```

Output:

```
Enter first number
10
Enter second number
3
Enter 1 for addition or 2 for multiplication
2
The result is 30
```

6. Comma Operator

Comma operator is a set of expression separated by comma is a valid constant in the C language. For example i and j are declared by the statements

```
Inti,j;
i=(j=10, j+20);
```

In the above declaration, right hand side consists of two expressions separated by comma. The first expression is j=10 and second is j+20. These expressions are evaluated from left to right. ie first the value 10 is assigned to j and then expression j+20 is evaluated so the final value of i will be 30.

C – COMMA OPERATOR EXAMPLE PROGRAM:

```
#include<stdio.h>
#include<conio.h>
void main()
{
inti,j;
clrscr();
i=(j=10,j+20); // here expressions are evaluated from left to right. ie first the value 10 is assigned to j and then expression j+20 is evaluated so the final value of i will be 30.
printf("i=%d\nj=%d\n",i,j );
getch();
}
```

Output :

```
i = 30
j = 10
```

7. Unary Operator

Unary operators are special operators that operate upon one operand only. n C, there are quite a few unary operators as mentioned below :-

- The increment and
- The decrement operator.

The increment operator increases the value of a variable by one. And it is of two types, the post increment and pre increment.

The pre increment operator first increases the value and then uses it for calculation whereas the post increment operator does the otherwise.

Similar explanation can be extended for the decrement operator which also has two type same as that of the increment operator.

Similarly, there are other unary operators in C such as the address-of operator denoted by the ampersand symbol '&' ,which returns the address of its operand. We have following Unary Operators in C.

! (logical negation),

~ (one's complement or bitwise negation)

– (unary minus),

+ (unary plus),

& (addressof),

* (dereferencing),

++ (pre-increment),

— (pre-decrement),

sizeof operator,

(type) or cast operator

All Unary Operators have *associativity from Right to Left.*

Example of sizeof() Operator

```
#include <stdio.h>

int main()
{
    int x=10;

    printf("size of x : %d\n",sizeof(x));
    printf("size of 10 : %d\n",sizeof(10));
    printf("size of int : %d\n",sizeof(int));
    printf("size of char : %d\n",sizeof(char));
    printf("size of float: %d\n",sizeof(float));
```

```
    return 0;
}
```

Output

```
size of x : 4
size of 10 : 4
size of int : 4
size of char : 1
size of float: 4
```

8. Bitwise Operator

These operators are used to perform bit operations. Decimal values are converted into binary values which are the sequence of bits and bit wise operators work on these bits.

Bit wise operators in C language are & (bitwise AND), | (bitwise OR), ~ (bitwise OR), ^ (XOR), << (left shift) and >> (right shift).

The following table lists the Bitwise operators supported by C. Assume variable 'A' holds 60 and variable 'B' holds 13, then –

Operator	Description	Example
&	Binary AND Operator copies a bit to the result if it exists in both operands.	(A & B) = 12, i.e., 0000 1100
|	Binary OR Operator copies a bit if it exists in either operand.	(A | B) = 61, i.e., 0011 1101
^	Binary XOR Operator copies the bit if it is set in one operand but not both.	(A ^ B) = 49, i.e., 0011 0001
~	Binary Ones Complement Operator is unary and has the effect of 'flipping' bits.	(~A) = -61, i.e,. 1100 0011 in 2's complement form.
<<	Binary Left Shift Operator. The left operands value is moved left by the number of bits specified by the right operand.	A << 2 = 240 i.e., 1111 0000
>>	Binary Right Shift Operator. The left operands value is moved right by the number of bits specified by the right operand.	A >> 2 = 15 i.e., 0000 1111

Following Example To Understand All The Bitwise Operators Available In C –

```
#include <stdio.h>

main() {

 unsigned int a = 60; /* 60 = 0011 1100 */
 unsigned int b = 13; /* 13 = 0000 1101 */
 int c = 0;

 c = a & b; /* 12 = 0000 1100 */
 printf("Line 1 - Value of c is %d\n", c );

 c = a | b; /* 61 = 0011 1101 */
 printf("Line 2 - Value of c is %d\n", c );

 c = a ^ b; /* 49 = 0011 0001 */
 printf("Line 3 - Value of c is %d\n", c );

 c = ~a; /*-61 = 1100 0011 */
 printf("Line 4 - Value of c is %d\n", c );

 c = a << 2; /* 240 = 1111 0000 */
 printf("Line 5 - Value of c is %d\n", c );

 c = a >> 2; /* 15 = 0000 1111 */
 printf("Line 6 - Value of c is %d\n", c );
}
```

When you compile and execute the above program, it produces the following result –

```
Line 1 - Value of c is 12
Line 2 - Value of c is 61
```

```
Line 3 - Value of c is 49
Line 4 - Value of c is -61
Line 5 - Value of c is 240
Line 6 - Value of c is 15
```

EXERCISE

Short Questions-2 marks (each)

1. How "C" language is used as Modular Programming.
2. What is the use of printf() function and what it returns?
3. Explain the function of header file #include < >. How it is different from #include " ".
4. What is the sizeof() int and char in C.
5. What is Ternary Operator. Explain with an example.

Long Questions (5 marks each)

1. In which year, the C language was developed How we can say that C is a user-friendly language over Assembly language. What are the applications of C language in practical scenario?
2. Explain variable, keyword, token and format specifiers with an example.
3. Write a program in C to print the ASCII value of a character.
4. Write the logic for whether the entered character is alphabet or not.
5. What are different types of Operators available in C. Explain in detail with an example.

Multiple Choice Questions

1. Choose the correct one

   ```
   int a= 5.5 + 4.5;
   ```

   a) a=9

   b) a=10

c) a=0

d) a=10.0

Answer: b

2. Choose the correct one

```
int a= 5.3 + 4.5;
```

a) a=9

b) a=10

c) a=0

d) a=10.0

Answer: a

3. Choose the correct one

```
int main()
{
int a =345%10;
printf("%d", a);
return 0;
}
```

a) a=8

b) a=15

c) a=5

d) a=10

Answer: c

4. Choose the correct one

```
#define DELF 16
int main()
{
int d=20;
d = d*DELF;
printf("DELF=%d",d);
return 0;
}
```

a) DELF=320

b) DELF=16

c) Compiler error

d) DELF=0

Answer: a

5. Choose the correct one

```
Sizeof(void) is
```

a) platform dependent

b) platform independent

c) both a & b

c) none of the above

Answer: a

6. Choose the correct one

```
#include <stdio.h>
int main()
{
int d;
printf("size of d= %d", sizeof(d));
return 0;
}
```

a) size of d=2

b) size of d=8

c) size of d=4

d) size of d=1

Answer: c

7. Choose the correct one

```
#include<stdio.h>
int main()
{
printf("%x\n", -1>>1);
return 0;
}
```

a) 0000

b) fffffff0

c) 0fffffff
d) ffffffff

Answer: d

8. Bitwise & and | are ____________ operators
   a) unary operators
   b) binary operators
   c) both a & b
   d) none of the above

Answer: b

9. Choose the correct one

```
#include<stdio.h>
int main()
{
int b=0;
b=6+4/2*6;
printf("value = %d", b);
return 0;
}
```

   a) value=8
   b) value=18
   c) value=10
   d) none of the above

Answer: b

10. Choose the correct one

```
#include<stdio.h>
int main()
{
int a = 6, b = 2, d;
d = (a == 6 ? (b == 4 ? 3 : 5) : 0);
printf ("%d\n", d);
return 0;
}
```

a) 6
b) 0
c) 4
d) 5

Answer: d

4

Control Structures

4.1 INTRODUCTION

Control structures in C programming are constructs that enable you to control the flow of program execution based on certain conditions. These structures allow you to specify the order in which statements are executed, based on specific conditions or criteria. There are three main types of control structures in C: selection, iteration, and jump statements.

- **Selection Statements**: Selection statements enable you to choose which section of code should be executed based on a certain condition. The two most commonly used selection statements in C are **the if statement** and the switch statement. The if statement allows you to execute a block of code if a certain condition is true, while the switch statement allows you to select from multiple options based on the value of a variable.
- **Iteration Statements**: Iteration statements, also known as loop statements, allow you to execute a block of code repeatedly as long as a certain condition is true. The three most commonly used iteration statements in C are the while loop, the do-while loop, and the for loop. The while loop allows you to execute a block of code as long as a certain condition is true, while the do-while loop is similar but will always execute the code at least once. The for loop allows you to specify an initial condition, a

loop condition, and an update statement to control the flow of the loop.

- **Jump Statements**: Jump statements allow you to transfer control of your program to a different section of code based on certain conditions. The three jump statements in C are break, continue, and goto. The break statement allows you to exit a loop or switch statement, while the continue statement allows you to skip over certain parts of a loop. The goto statement allows you to transfer control to a different part of your program, but it should be used with caution as it can make your code difficult to read and maintain.

Overall, control structures in C are essential tools for programming and can help you to create efficient and well-structured code. By using the right control structures in the right situations, you can create programs that are flexible and easily adaptable to different scenarios.

In C programming "selection construct" or "conditional statement" is used for decision making. Diagram 1 illustrates "selection construct".

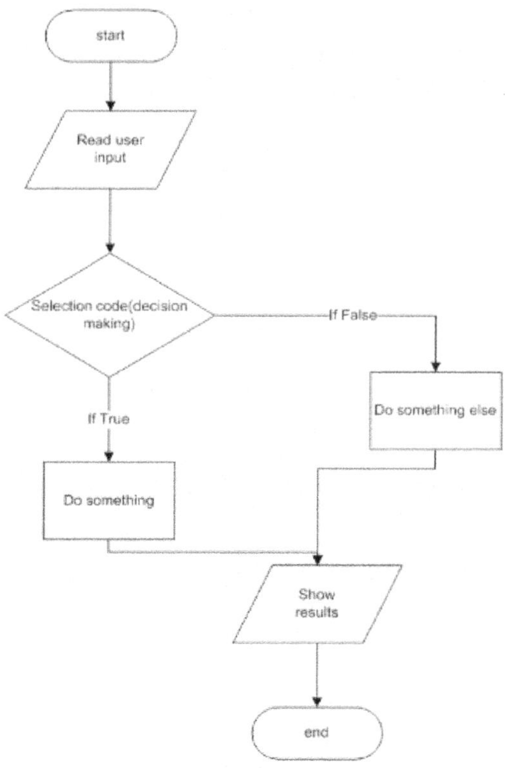

Diagram 1: Simple selection construct

Conditional statement is the term used by many programming languages. The importance of conditional statements should not be ignored because every program has an example of these statements. "IF statement" and "switch statement" are the most popular conditional statements used in C.

We can perform it by 4 statement:-

1. if
2. switch
3. conditional operator statement
4. Go to statement

IF Statement

`if` statement evaluates the test expression inside the parenthesis by this way :-

- If the test expression is evaluated to true (nonzero), statements inside the body of `if` is executed.
- If the test expression is evaluated to false (0), statements inside the body of `if` is skipped from execution.

It is used by 4 ways:-

- Simple if
- if else
- Nested if
- Else if ladder

1. Simple IF

No Else part

Syntax:

```
if(condition)
{
 statements;
}
```

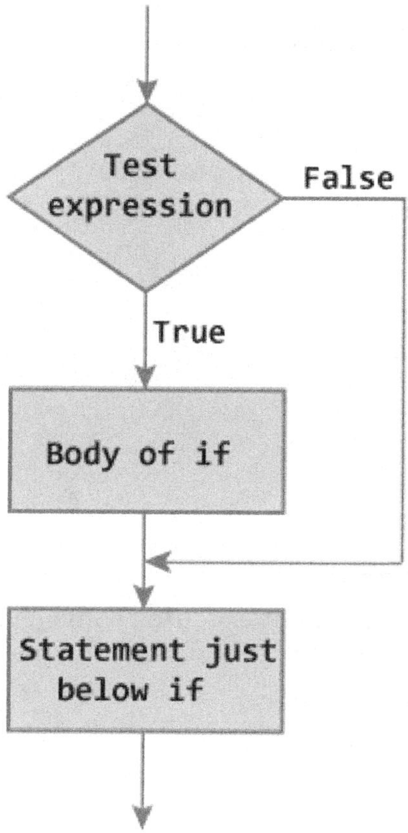

Figure: Flowchart of if Statement

Programme for Even No :- if even then print only

```
#include<stdio.h>
#include<conio.h>
void main()
{
int i;
clrscr();
printf("Enter the number");
scanf("%d",&i);
if(i%2==0)
```

```
{
printf("No. is even");
}
printf("Value of %d",i);
getch();
}
```

Output

```
Enter the no:- 5
value of i is 5
Enter the no. : 8
even no. value of 8
```

2) if else statement

The `if...else` statement executes some code if the test expression is true (nonzero) and some other code if the test expression is false (0).

Syntax:

```
    If (condition)
    {
    Body
    }
Else
{
Body
}
```

if test expression is true, codes inside the body of `if` statement is executed and, codes inside the body of `else` statement will not execute.

If test expression is false, codes inside the body of `else` statement is executed and, codes inside the body of `if` statement will not execute.

Entry Point

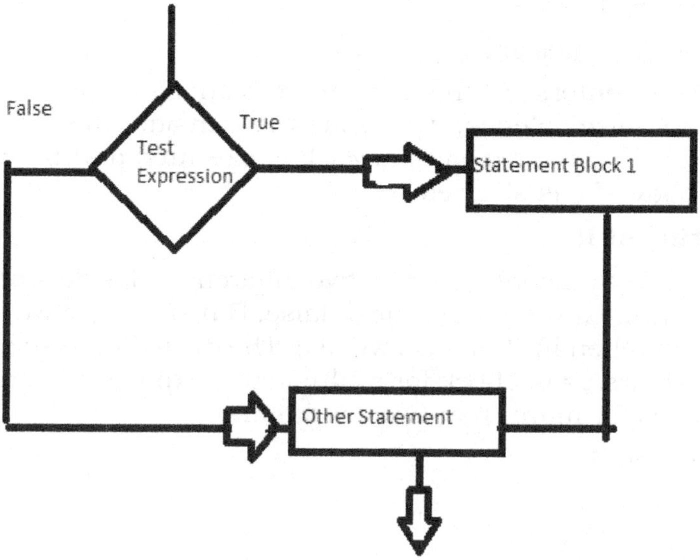

Figure: Flow Chart of if else statement

Program to check whether a no. is even or odd.

```
#include<stdio.h>
#include<conio.h>
Void main()
{
int i;
Clrscr();
Printf("Enter a number");
Scanf("%d",&i);
If(i%2==0)
Printf("%d is even number",i);
Else
Printf("%d is odd number",i);
getch();
}
```

Output

```
Enter an number: 19
19 is an odd number.
```

When user enters 7, the test expression (number%2 == 0) is evaluated to false. Hence, the statement inside the body of else statement printf("%d is an odd integer"); is executed and the statement inside the body of if is skipped.

3) Nesting of if

The if...else statement executes two different codes depending upon whether the test expression is true or false. But sometime we have more than 2 options then in that case we use .The nested if...else statement. Because it allows us to check for multiple test expressions and execute different codes for more than two conditions.

```
If(condtion 1)
{
If(condition 2)
{
Block 1
}
Else
{
Block 2
}
}
Else
{
Block 3
}
```

Program to print greatest value from 3 values using nested if else statement.

```
#include<stdio.h>
#include<conio.h>
Void main()
```

```
{
int i,j,k;
Printf("Enter due value of i,j,k");
Scanf("%d%d%d,&i,&j,&k");
If(a>b)
{
If(a>c)
Printf("a is greatest");
Else
Printf(("c is greatest");
}
Else
{
If(b>c)
Printf("b is greatest%d");
Else
Printf("c is greatest no. %d",c);
}
getch();
}
```

Output

```
Enter value of i , j, k = 2,6,7
7 is greatest number.
```

4) Else If ladder

The else if ladder is a way of putting multiple ifs together when multipath decisions are involved. It is a one of the types of decision making and branching statements. A multipath decision is a chain of if's in which the statement associated with each else is an if.

Syntax:

```
If(condition 1)
{
```

```
Block 1
}
Else if (Condition 2)
{
Block 2
}
Else if(Condition 3)
{
Block 3
}
Else
{
Default Block
}
```

This construct is known as the else if ladder. The conditions are evaluated from the top of the ladder to downwards. As soon as a true condition is found, the statement associated with it is executed and the control is transferred to the statement-x (skipping the rest of the ladder). When all the n conditions become false, then the final else containing the default statement will be executed.

Here's an example of an elseif ladder in C:

```
#include <stdio.h>
int main()
{
 int num;
 printf("Enter a number: ");
 scanf("%d", &num);
 if (num == 0)
 {
 printf("You entered zero.\n");
 }
else if (num < 0)
{
```

```
 printf("You entered a negative number.\n");
 }
 else if (num % 2 == 0)
 {
 printf("You entered an even number.\n");
 }
else
 {
 printf("You entered an odd number.\n");
 }
 return 0;
}
```

Output or various input values:

```
Enter a number: 0
You entered zero.

Enter a number: -5
You entered a negative number.

Enter a number: 12
You entered an even number.

Enter a number: 3
You entered an odd number.
```

In this example, the user is prompted to enter a number, and the program uses an elseif ladder to determine whether the number is zero, negative, even, or odd. Based on the input, the program outputs a corresponding message.

4.2 LOOP CONTROL STATEMENTS

- Loop control statements in C are used to repeat a specific block until some end condition is satisfied.

- A program loop is made up of two parts one part is known as body of the loop and the other is known as control condition.
- Depending on the control condition statement the statements within the loop may be executed repeatedly.
- Depending on the position of the control statement in the loop, a control structure may be classified either as an entry controlled loop or as an exit controlled loop.

Entry Controlled Loop

- When the control statement is placed before the body of the loop then such loops are called as entry controlled loops.
- If the test condition in the control statement is true then only the body of the loop is executed.
- If the test condition in the control statement is not true then the body of the loop will not be executed. If the test condition fails in the first checking itself the body of the loop will never be executed.

Exit Controlled Loop

- When the control statement is placed after the body of the loop then such loops are called as exit controlled loop.
- In this the body of the loop is executed first then the test condition in the control statement is checked.
- If it is true then the body of the loop is executed again.
- If the test condition is false, the body of the loop will not be executed again. In exit controlled loops even if the test condition fails in the first attempt itself the body of the loop is executed at least once.

Following are the steps in a looping process:

- Initialization of loop control variable
- Test of a specific condition for the execution of the loop
- Execution of the statements in the body of the loop
- Altering the value of the loop control variable

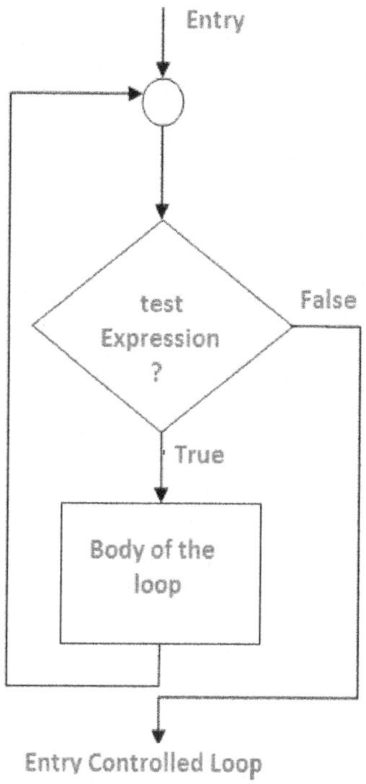

Figure (a): Flow chart of Entry Controlled loop

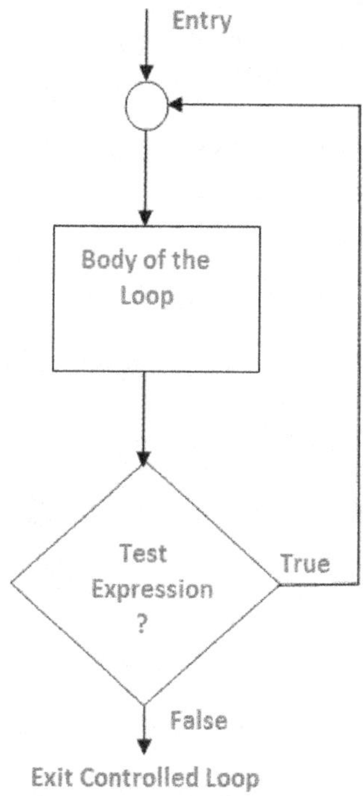

Figure (b): Flow chart of Exit Controlled loop

In the beginning the loop control variable is initialized after that step2, step3 and step4 are carried out till specified test condition becomes false.

The test may be either to determine whether the loop has been repeated the specified number of times or to determine whether a particular condition has been met. The C language provides 3 loop structures.

There are 3 types of loop control statements . They are,

- while
- do-while
- For

1. While loop

It is entry control loop first it checked the condition then only it execute the body of loop.

Syntax

```
While( condition)
{
Body of loop
}
```

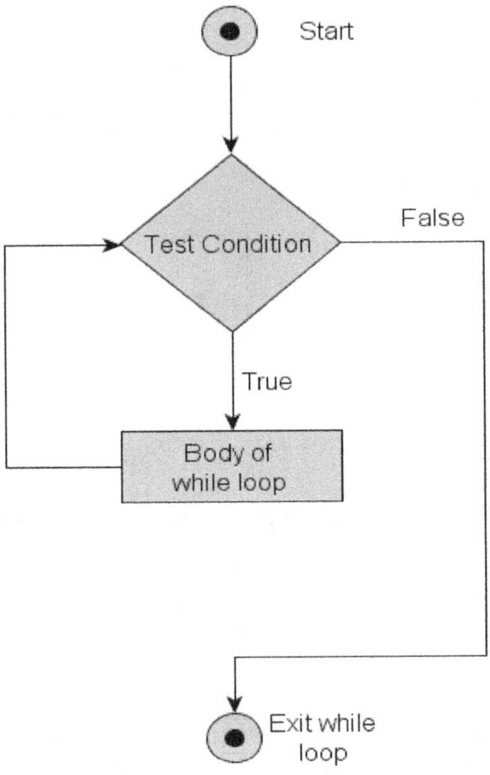

Figure: Flow Chart of While Loop

```
#include<stdio.h>
#include<conio.h>
Void main()
{
int i=1;
clrscr();
while(i<=5)
```

```
{
Printf(" print number is %d",i);
i++;
}
getch();
}
```

Output

```
Number is 1
Number is 2
Number is 3
Number is 4
Number is 5
```

- **Step1:** The variable count is initialized with value 1 and then it has been tested for the condition.
- **Step2:** If the condition returns true then the statements inside the body of while loop are executed else control comes out of the loop.
- **Step3:** The value of count is incremented using ++ operator then it has been tested again for the loop condition.

2. Do While Loop

It is exit control loop first it execute the body of loop then checked the condition only.

Syntax

```
do
{
Body of loop
}while (test condition);
```

Program for even number between 1 to 10 using do while.

```
#include<stdio.h>
#include<conio.h>
```

```
Void main()
{
int i=2;
clrscr();
do
{
Printf(" print number is %d",i);
I=i+2;
} while(i<=10)

getch();
}
```

Output

```
Number is 2
Number is 4
Number is 6
Number is 8
Number is 10
```

3. FOR Loop

for loop is an entry controlled loop i.e. the condition is checked before entering into the loop. So if the condition is false for the first time, the statements inside loop may not be executed at all. In order to exit from a for loop, either the condition should be false or a break statement should be encountered.

Syntax

```
For( initialization; condition; increment/decrement)
{
Body of loop;
}
```

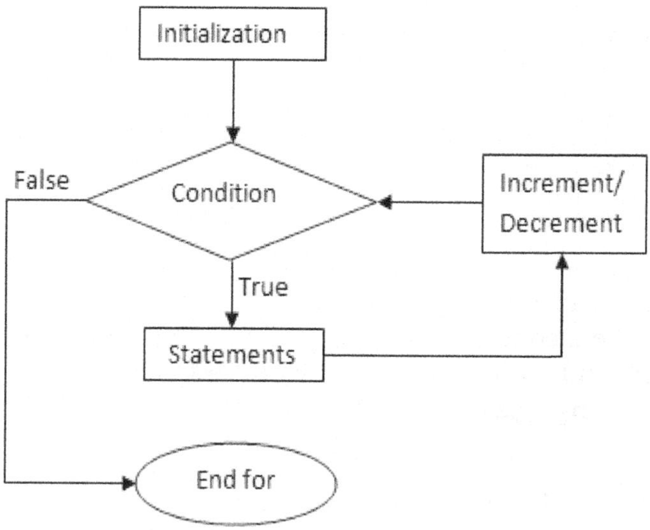

fig: Flowchart for for loop

Figure: Flow Chart for the loop

Program for odd number between 1 to 10 using for loop.

```
#include<stdio.h>
#include<conio.h>
Void main()
{
int i;
clrscr();
for(i=1;i<=10;i+2)
{
Printf(" print number is %d",i);
}
getch();
}
```

Output

```
Number is 1
Number is 3
Number is 5
Number is 7
Number is 9
```

Infinite for Loop

There may be a condition in a for loop which is always true. In such case, the loop will run infinite times. **For example**,

```
for (i=0; i>0; i++)
{
 printf("This is infinite loop");
}
```

In the example, program execution checks the condition *i > 0* which always holds true. It makes it run infinite times printing the underlying statement *This is infinite loop.*

Infinite Loop with No Conditions

Likewise, when the condition part is left blank, the C compiler will also treat it as *true* and the loop will run infinite times.

```
for (i=0; ; i++)
{
 printf("This is infinite loop");
}
for (;;)
{
 printf("This is infinite loop");
}
```

Break Statements in Infinite Loop

We can use break statement to exit from an infinite loop. **For example**,

```
for (;;)
```

```
{
 printf("This loop will run only once");
 break;
}
```

4.3 SWITCH STATEMENT

It is used to solve multiple option type problems, or we can say it is alternate to long if statements. for example: Menu like program, where one value is associated with each option and you need to choose only one at a time, then, switch statement is used.

The expression in switch evaluates to return an integral value, which is then compared to the values present in different cases. It executes that block of code which matches the case value. If there is no match, then **default** block is executed(if present). The general form of switch statement is,

```
switch(expression)
{
 case (value-1):
   block-1;
   break;
 case (value-2):
   block-2;
   break;
 case (value-3):
   block-3;
   break;

 default:
    default-block;
   break;
}
```

Rules for using `switch` statement

The expression (after switch keyword) must yield an **integer** value i.e the expression should be an integer or a variable or an expression that evaluates to an integer.

- The case **label** values must be unique.
- The case label must end with a colon(:)
- The next line, after the **case** statement, can be any valid C statement.

Points to Remember

Don't use those expressions to evaluate switch case, which may return floating point values or strings or characters.

break statements are used to **exit** the switch block. Use of break statement is optional after each block, but if do not use it, then all the consecutive blocks of code will get executed after the matching block.

For example :-

```
inti=1;
switch(i)
{

case 1:
printf("A"); // No break
case 2:
printf("B"); // No break
case 3:
printf("C");
break;
}
```

```
A B C
```

The output was supposed to be only **A** because only the first case matches, but as there is no break statement after that block, the next blocks are executed too, until it a break statement in encountered or the execution reaches the end of the switch block.

default case is executed when none of the mentioned case matches the switch expression. The default case can be placed anywhere in the switch case. Even if we don't include the default case, switch statement works.

Nesting of switch statements are allowed, which means you can have switch statements inside another switch block. However, nested switch statements should be avoided as it makes the program more complex and less readable.

A program to demonstrate switch statement.

```
#include <stdio.h>
int main()
{
 int x = 3;
 switch (x)
 {
 case 1: printf("Choice is 1");
 break;
 case 2: printf("Choice is 2");
 break;
 case 3: printf("Choice is 3");
 break;
 default: printf("Choice other than 1, 2 and 3");
 break;
 }
 return 0;
}
```

Output:

```
Choice is 3
```

4.4 BREAK OPERATOR

It is used to come out of the loop instantly. When a break statement is encountered inside a loop, the control directly comes out of loop and the loop gets terminated. It is used with if statement, for ,while,do while loop.

This can also be used in switch case control structure. Whenever it is encountered in switch-case block, the control comes out of the switch-case.

Syntax

```
Break;
```

Flow chart of break statement

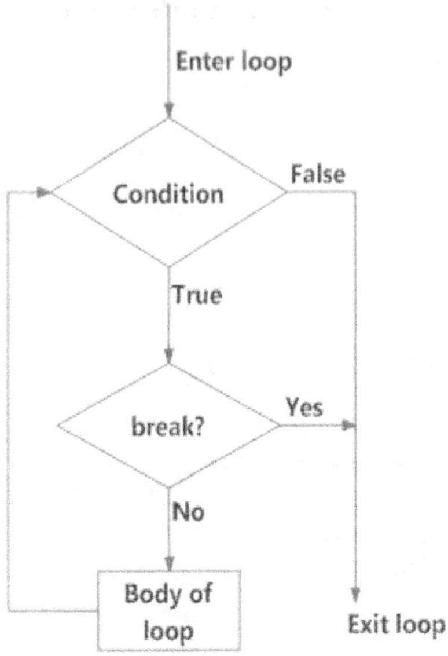

Figure: Flowchart of Break

Program to demonstrate use of break in a for loop

```
#include <stdio.h>
int main()
{
 int var;
 for (var =100; var>=10; var --)
 {
 printf("var: %d\n", var);
 if (var==99)
 {
 break;
 }
 }
 printf("Out of for-loop");
 return 0;
}
```

Output:

```
var: 100
var: 99
Out of for-loop
```

4.5 CONTINUE OPERATOR

The continue operator forces the loop to continue or execute the next iteration. When the continue statement is executed in the loop, the code inside the loop following the continue statement will be skipped and next iteration of the loop will begin.

It's used when we realized (or the code does) that this iteration of the loop isn't going to produce a useful result

Syntax

Continue;

Flow chart of continue operator.

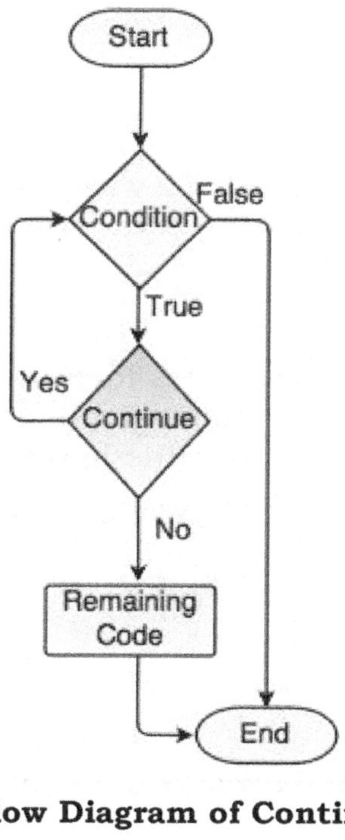

Figure: Flow Diagram of Continue Statement

Program to demonstrate the use of Continue operator

```
#include<stdio.h>
int main()
{
int i;
for(i=0;i<5;i++)
{
if(i==3)
{
continue;
}
printf("%d",i\t);
}
```

Output :

```
0 1 2 4
```

4.6 COMMA OPERATOR

Comma represented by , is used in 2 way thats is either as operator or as separator . we will discuss here both form one by one .

First Comma as operator

It is a binary operator that evaluates its first operand and discards the result, it then evaluates the second operand and returns this value (and type). The comma operator has the lowest precedence of any C operator, and acts as a sequence point.

```
/* comma as an operator */
int i = (75, 100); /* 100 is assigned to i*/
```

Second way Comma as separator

While declaration multiple variables and providing multiple arguments in a function, comma works as a separator.

Example:

```
int a,b,c;
```

In this statement, **comma is a separator and tells to the compiler that these (a, b, and c) are three different variables**.

Goto Statements

This statement is used to transfer the control on specific point in program. By GOTO we can transfer either in forward or backward direction.

For this we need label & label is written with semi colon.

Label: // it identify where to transfer the program control.

Fig shows forward and backward jump using label in goto statement.

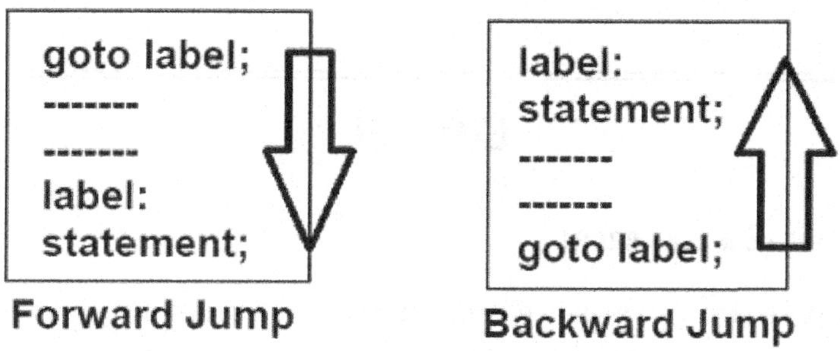

Figure a) Forward Jump **Figure b) Backward Jump**

Program to check whether number is odd or even using GOTO statement.

```
# include<stdio.h>
# include<conio.h>
int main()
{
 int sum=0;
 for(int i = 0; i<=10; i++){
   sum = sum+i;
   if(i==5){
    goto addition;
   }
 }
 addition:
 printf("%d", sum);
```

```
  return 0;
}
```

Output:

```
15
```

Explanation: In this example, we have a label addition and when the value of i (inside loop) is equal to 5 then we are jumping to this label using goto. This is reason the sum is displaying the sum of numbers till 5 even though the loop is set to run from 0 to 10.

EXERCISE

Short questions (2 marks each)

1. What is the purpose of a decision statement in programming?
2. How many expressions are evaluated in an if statement?
3. Which loop control statement will always execute at least once?
4. What is the purpose of a break statement in a loop?
5. Which jump statement should be avoided in modern programming?

Long questions (5 marks each)

1. What is the difference between an if statement and a switch statement? Provide an example for each.
2. Explain the three parts of a for loop and give an example of a for loop in action.
3. What is the difference between a while loop and a do-while loop? Provide an example for each.
4. What is the purpose of a break statement in a loop? How can it be used to optimize code?
5. Discuss the drawbacks of using a goto statement in programming. Provide an example of code that can be written without using a goto statement.

Multiple choice questions

1. What is the output of the following code?

```
int x = 3;
switch (x) {
 case 1:
 System.out.println("Case 1");
 break;
 case 2:
 System.out.println("Case 2");
 break;
 default:
 System.out.println("Default case");
}
```

   a) Case 1
   b) Case 2
   c) Default case
   d) No output

Answer: c) Default case

2. What is the output of the following code?

```
int i = 0;
while (i < 5) {
 System.out.println(i);
 i++;
}
```

   a) 0 1 2 3 4
   b) 1 2 3 4 5
   c) 0 1 2 3 4 5
   d) No output

Answer: a) 0 1 2 3 4

3. What is the output of the following code?

```
for (int i = 0; i < 5; i++) {
 if (i == 2) {
```

```
    continue;
    }
    System.out.println(i);
  }
```

a) 0 1 2 3 4

b) 0 1 3 4

c) 2 3 4

d) No output

Answer: b) 0 1 3 4

4. What is the output of the following code?

```
int i = 0;
do {
  System.out.println(i);
  i++;
} while (i < 5);
```

a) 0 1 2 3 4

b) 1 2 3 4 5

c) 0 1 2 3 4 5

d) No output

Answer: a) 0 1 2 3 4

5. What is the output of the following code?

```
for (int i = 0; i < 5; i++) {
  if (i == 2) {
  break;
  }
  System.out.println(i);
}
```

a) 0 1 2 3 4

b) 0 1

c) 0 1 2

d) No output

Answer: b) 0 1

5

Arrays

5.1 INTRODUCTION

Arrays are a fundamental data structure in computer science and are used to store collections of data. They are used to store a sequence of values of the same data type, where each value can be accessed using an index or subscript. In C programming, arrays are a powerful tool that can be used to store large amounts of data and perform operations on that data in a fast and efficient manner.

Syntax:

The syntax for declaring an array in C is as follows:

```
data_type array_name[array_size];
```

Here, data_type is the type of data that the array will store, array_name is the name given to the array, and array_size is the number of elements that the array will hold.

For example, to declare an array numbers that will store 5 integers, we would use the following syntax:

```
int numbers[5];
```

Code:

Arrays can be initialized at the time of declaration by specifying the values of the elements. The syntax for initializing an array is as follows:

```
data_type array_name[array_size] = {element1, element2, ...,
element_array_size};
```

For example, to declare and initialize an array numbers with the values 1, 2, 3, 4, and 5, we would use the following code:

```
int numbers[5] = {1, 2, 3, 4, 5};
```

To access the elements of an array, we use the array name followed by the index of the element in square brackets. The index of the first element is 0, the second element is 1, and so on.

For example, to access the third element of the numbers array, we would use the following code:

```
printf("The third element of the numbers array is: %d", numbers[2]);
```

Example:

Let's see an example that demonstrates the use of arrays in C programming. The following code accepts 5 numbers from the user and stores them in an array, then prints the sum of these numbers.

```
#include <stdio.h>
int main() {
 int numbers[5];
 int sum = 0;

 printf("Enter 5 numbers: ");
 for (int i = 0; i < 5; i++) {
 scanf("%d", &numbers[i]);
 sum += numbers[i];
 }
 printf("The sum of the numbers is: %d", sum);
 return 0;
}
```

In this example, we first declared an array numbers that will store 5 integers. Then, we used a for loop to accept 5 numbers from the user and store them in the numbers array. Finally, we calculated the sum of these numbers and printed it to the console.

There are two main types of arrays in C programming: one-dimensional arrays and multi-dimensional arrays. In this chapter, we will be first focusing on one-dimensional arrays.

5.2 ONE-DIMENSIONAL ARRAYS

A one-dimensional array is a sequence of elements of the same data type stored in contiguous memory locations. The elements in the array are indexed from 0 to n-1, where n is the number of elements in the array. In C programming, one-dimensional arrays are declared by specifying the data type, followed by the array name and the size of the array in square brackets.

Syntax:

```
data_type array_name[array_size];
```

For example, the following code declares an array of 10 integers:

```
int numbers[10];
```

Accessing Elements in a One-Dimensional Array

Once you have declared a one-dimensional array, you can access individual elements by using the array name followed by the index in square brackets. The index must be an integer value between 0 and n-1, where n is the number of elements in the array.

Syntax:

```
array_name[index];
```

For example, the following code assigns the value 100 to the first element of the numbers array:

```
numbers[0] = 100;
```

Initializing a One-Dimensional Array

You can initialize a one-dimensional array at the time of declaration by specifying the values of the elements separated by commas and enclosed in curly braces.

Syntax:

```
data_type array_name[] = {element1, element2, ..., elementn};
```

For example, the following code declares and initializes an array of 10 integers:

```
int numbers[] = {1, 2, 3, 4, 5, 6, 7, 8, 9, 10};
```

Example: Summing the Elements of a One-Dimensional Array

The following code is an example of how to use a one-dimensional array in C programming. The program calculates the sum of the elements in an array of 10 integers.

```
#include <stdio.h>
int main()
{
 int numbers[10] = {1, 2, 3, 4, 5, 6, 7, 8, 9, 10};
 int sum = 0;
 int i;

 for (i = 0; i < 10; i++)
 {
 sum = sum + numbers[i];
 }

 printf("The sum of the elements in the array is: %d\n",
sum);

 return 0;
}
```

Output:

```
The sum of the elements in the array is: 55
```

In this example, we declare an array of 10 integers and initialize it with the values 1 to 10. We then use a for loop to iterate through the elements of the array and calculate the sum of element of array .

5.3 MULTI-DIMENSIONAL ARRAYS IN C

In C programming, a single dimensional array can be thought of as a list of elements of the same data type. However, in real-world scenarios, data can be organized in more complex structures, such as matrices, tables, and grids. Multi-dimensional arrays provide a way to represent and manipulate such data structures in C. A multi-dimensional array is essentially an array of arrays.

Syntax

A multi-dimensional array in C is declared as follows:

```
data_type array_name[size1][size2]...[sizeN];
```

Where data_type is the type of data the array will store (e.g. int, float, char, etc.), array_name is the name given to the array, and size1, size2,

..., sizeN are the sizes of the dimensions of the array. For example, the following code declares a 2-dimensional integer array with 4 rows and 3 columns:

```
int arr[4][3];
```

Accessing Array Elements

To access the elements of a multi-dimensional array, we use the array name followed by the indices of the elements in square brackets. For a 2-dimensional array, the syntax would look like this:

```
array_name[row][column];
```

For example, the following code accesses the element in the second row and third column of the array arr declared above:

```
arr[1][2];
```

Initializing Multi-Dimensional Arrays

Multi-dimensional arrays can be initialized at the time of declaration just like single dimensional arrays. The syntax is similar, but with multiple sets of values enclosed in curly braces {} to represent the elements in each dimension.

Here's an example of initializing a 2-dimensional integer array with 4 rows and 3 columns:

```
int arr[4][3] = {{1, 2, 3}, {4, 5, 6}, {7, 8, 9}, {10, 11, 12}};
```

Note that it's possible to leave out the size of some dimensions in the initialization, and C will automatically deduce the size based on the number of elements provided. For example:

```
int arr[][3] = {{1, 2, 3}, {4, 5, 6}, {7, 8, 9}, {10, 11, 12}};
```

This declaration is equivalent to the previous one.

Traversing a Multi-Dimensional Array

To traverse a multi-dimensional array, we use nested loops, where the outer loop represents the rows and the inner loop represents the columns. The following code demonstrates how to traverse a 2-dimensional integer array:

```
int arr[4][3] = {{1, 2, 3}, {4, 5, 6}, {7, 8, 9}, {10, 11, 12}};
```

```
int i, j;
for (i = 0; i < 4; i++) {
 for (j = 0; j < 3; j++) {
 printf("%d ", arr[i][j]);
 }
 printf("\n");
}
```

This code will print the following output:

```
1 2 3
4 5 6
7 8 9
10 11
```

EXERCISE

Short Questions (2 Marks)

1. What is an array?
2. What is the difference between one-dimensional and two-dimensional arrays?
3. What is the syntax for declaring a one-dimensional array in C?
4. How are multi-dimensional arrays stored in memory?
5. How can you access an element in a two-dimensional array?

Long Questions (5 Marks)

1. What are the advantages of using arrays in programming? Give examples.
2. Explain the process of declaring and initializing a one-dimensional array in C with suitable examples.
3. How are two-dimensional arrays declared and initialized in C? Explain with an example.

4. Discuss the differences between static and dynamic memory allocation of arrays.
5. How can you pass a one-dimensional array to a function in C? Explain with an example.

Multiple Choice Questions

1. Which of the following is not a valid data type for an array in C?
   a) int
   b) float
   c) char
   d) string

Answer: d

2. What is the index of the first element in a one-dimensional array?
   a) 0
   b) 1
   c) -1
   d) none of the above

Answer: a

3. How many indices are required to access an element in a two-dimensional array?
   a) 1
   b) 2
   c) 3
   d) it depends on the size of the array

Answer: b

4. What is the maximum number of dimensions an array can have in C?
   a) 1
   b) 2
   c) 3
   d) there is no limit

Answer: d

5. Which of the following is a valid way to initialize a one-dimensional array in C?
   a) int arr[5] = {1, 2, 3}
   b) int arr[] = {1, 2, 3}
   c) int arr[3] = {1, 2, 3, 4, 5}
   d) int arr[] = 1, 2, 3

Answer: b

6. What is the output of the following code?

```
int arr[2][2] = { {1, 2}, {3, 4} };
printf("%d", arr[1][0]);
```

   a) 1
   b) 2
   c) 3
   d) 4

Answer: c

7. What is the difference between a static and a dynamic array?
   a) A static array has a fixed size, while a dynamic array can be resized during runtime.
   b) A static array is declared using the "static" keyword, while a dynamic array is not.
   c) There is no difference between a static and a dynamic array.
   d) A static array is stored in the stack, while a dynamic array is stored in the heap.

Answer: a

8. How can you pass a two-dimensional array to a function in C?
   a) By passing each element of the array individually.
   b) By passing a pointer to the first element of the array.
   c) By passing a pointer to the entire array.
   d) It is not possible to pass a two-dimensional array to a function in C.

Answer: b

9. What is the size of a two-dimensional array with 3 rows and 4 columns, where each element is an integer?
   a) 3
   b) 4
   c) 12
   d) 24

Answer: c

10. What is the difference between an array and a pointer in C?
    a) An array can store a fixed size of elements, while a pointer can point to a single element or an array of elements.
    b) An array requires memory allocation at the time of declaration, while a pointer can be dynamically allocated memory during runtime.
    c) An array name can be used as a constant pointer to the first element of the array, while a pointer needs to be explicitly dereferenced to access the value it points to.

Answer: b) An array can only store one type of data, while a pointer can store any type of data.

6

Functions

Functions are the building blocks of a C program. They are a way of grouping together a set of instructions that perform a specific task. Functions help to break down a large program into smaller, more manageable, and reusable parts.

The **syntax** of a function in C is as follows:

```
return_type function_name( parameter_list )
{
 body of the function
}
```

Here, **return_type** is the data type of the value returned by the function. If the function does not return a value, the return_type should be void.

- **function_name** is the name of the function, which must follow the same naming conventions as variables in C.
- **parameter_list** is a list of parameters that the function accepts. Each parameter is a variable of a specific data type, separated by commas. If the function does not require any parameters, the parameter_list should be left empty.

The body of the function contains a series of statements that perform the desired task.

Let's look at an example of a function that calculates the factorial of a given number:

```
#include <stdio.h>
int factorial(int n)
{
 int i, result = 1;
 for(i=1; i<=n; i++)
 {
 result = result * i;
 }
 return result;
}
int main()
{
 int num;
 printf("Enter a number: ");
 scanf("%d", &num);
 printf("The factorial of %d is %d\n", num, factorial(num));
 return 0;
}
```

In this example, we have defined a function factorial that takes an integer n as an input and returns its factorial value. The function uses a loop to calculate the factorial and stores the result in the variable result. Finally, the function returns the result.

The main function is where the program starts executing. It prompts the user to enter a number and stores the value in the variable num. The factorial of num is then calculated by calling the factorial function and passing num as an argument. The result is then printed to the screen.

There are **two types of** functions in C:

1. **Built-in Functions**: These are predefined functions in C that are stored in the standard library. Examples of library functions are printf, scanf, strlen, etc.
2. **User-defined Functions**: These are functions that are defined by the user in their program. In the example above, the function factorial is a user-defined function.

In conclusion, functions are an important part of C programming and help to make the code more organized and easier to manage. By breaking down a program into smaller, reusable parts, functions allow for more efficient and effective coding.

6.1 BUILT-IN FUNCTIONS IN C PROGRAMMING

C programming is a popular high-level language that is widely used for developing software applications. One of the reasons why it is so popular is its rich library of built-in functions. These functions provide a quick and easy way to perform common tasks and reduce the amount of code that you need to write. In this chapter, we will explore the syntax and usage of some of the most commonly used built-in functions in C programming.

The printf() Function:

The printf() function is used to display output to the console. The function takes one or more arguments, which can be variables, constant values, or expressions. The arguments are passed to the function inside the parentheses and are separated by commas.

The **syntax** of the printf() function is:

```
printf("format string", list of arguments);
```

The "format string" is a string that specifies the format in which the arguments should be displayed. For example, you can use the %d format specifier to display an integer value, %f for a floating-point value, and %s for a string value.

Here is an example of using the printf() function to display the values of two variables:

```
#include <stdio.h>

int main() {
 int a = 10;
 float b = 20.5;
 printf("The value of a is %d and the value of b is %f\n",
a, b);
 return 0;
}
```

When you run this code, you will see the following output:

The value of a is 10 and the value of b is 20.500000

The scanf() Function

The scanf() function is used to read input from the user. The function takes one or more arguments, which are the address of the variables that will store the input values.

The **syntax** of the scanf() function is:

```
scanf("format string", list of arguments);
```

The "format string" is a string that specifies the format of the input. For example, you can use the %d format specifier to read an integer value, %f for a floating-point value, and %s for a string value.

Here is an example of using the scanf() function to read the values of two variables:

```
#include <stdio.h>
int main() {
 int a;
 float b;
 printf("Enter the value of a: ");
 scanf("%d", &a);
 printf("Enter the value of b: ");
 scanf("%f", &b);
 printf("The value of a is %d and the value of b is %f\n", a, b);
 return 0;
}
```

When you run this code, you will be prompted to enter the values of the variables:

Enter the value of a: 10

Enter the value of b: 20.5

The value of a is 10 and the value of b is 20.500000

User-Defined Functions in C Programming

User-defined functions allow programmers to break down a program into smaller, more manageable parts. This makes code easier to read, understand, and maintain.To create a user-defined function, the programmer must first declare the function by giving it a name and specifying the input parameters (if any) that the function will accept. The programmer then writes the code that will be executed when the function is called.Once the function is defined, it can be called from other parts of the program, allowing the programmer to reuse the same code multiple times without having to rewrite it each time..

Syntax for User-Defined Functions:

A user-defined function in C programming has the following syntax:

```
return_type function_name(parameter1, parameter2, ...)
 {
 // function body
}
```

Here,

return_type is the data type of the value the function returns. Some functions perform the required operations without returning a value. In this case, the return type is the keyword void.

function_name is the name given to the function. It should be meaningful and descriptive.

The parameters parameter1, parameter2, ... are the inputs to the function. These parameters are optional and a function may have no parameters.

Example:

Consider a simple program to find the factorial of a number. Here, we can create a user-defined function to calculate the factorial.

```
#include <stdio.h>
int factorial(int n)
 {
 int result = 1, i;
 for (i = 1; i <= n; i++) {
 result = result * i;
 }
 return result;
}

int main() {
 int num, result;
 printf("Enter a number: ");
```

```
    scanf("%d", &num);
    result = factorial(num);
    printf("The factorial of %d is %d", num, result);
    return 0;
}
```

In this example, we have defined a function factorial that calculates the factorial of a given number. The function takes an integer n as input and returns the result as an integer.

Code Output:

When we run the above code, the output will be:

Enter a number: 5

The factorial of 5 is 120

Detailed Explanation:

- The function factorial takes an integer n as input and returns an integer.
- The function initializes a variable result to 1 and a loop variable i to 1.
- The loop continues until i is less than or equal to n.
- In each iteration of the loop, result is multiplied by i and i is incremented.
- Finally, the value of result is returned.

In the main function, we take input from the user and call the function factorial with the input value. The returned value is stored in a variable result and is displayed on the screen.

6.2 STORAGE CLASSES

In C programming, storage classes play an important role in the life cycle of a variable. Storage classes determine the scope, visibility, and lifetime of a variable. A storage class defines the type of memory storage that should be allocated for a variable, and how it should be used throughout the program.

There are four storage classes in C programming:

1. Auto
2. Register
3. Static
4. External

In this chapter, we will take a detailed look at each of these storage classes, their syntax, and how they are used in C programming.

Auto Storage Class

The Auto storage class is the default storage class for all local variables in a function. Variables declared within a function with the auto keyword or without any keyword are considered as auto variables. The scope of auto variables is limited to the function in which they are declared. They are destroyed at the end of the function, and their value is not preserved between function calls.

Syntax:

```
auto data_type variable_name;
```

Example:

```
#include<stdio.h>
void auto_storage()
{
 auto int a = 10;
 printf("Value of a: %d\n", a);
}
int main()
{
 auto_storage();
 return 0;
}
```

Output:

```
Value of a: 10
```

Register Storage Class

The Register storage class is used to declare variables that should be stored in the CPU register. This storage class is used to optimize the program execution time by reducing the time required to access variables. However, the number of variables that can be stored in the CPU register is limited, so it's not guaranteed that a variable declared as register will be stored in the CPU register. The scope of the register variables is limited to the function in which they are declared.

Syntax:

```
register data_type variable_name;
```

Example:

```
#include<stdio.h>
void register_storage()
{
 register int a = 10;
 printf("Value of a: %d\n", a);
}
int main()
{
 register_storage();
 return 0;
}
```

Output:

```
Value of a: 10
```

Static Storage Class

The Static storage class is used to declare variables that have a local scope but their value is preserved between function calls. This means that the value of the static variable is not lost when the function is exited and is available the next time the function is called. The static variables are initialized only and their value is retained between function calls.

Syntax:

```
static data_type variable_name;
```

Example:

```
#include<stdio.h>
void static_storage()
{
 static int a = 10;
 printf("Value of a: %d\n", a);
 a++;
}
```

```
int main()
{
 static_storage();
 static_storage();
 static_storage();
 return 0;
}
```

Output:

```
Value of a: 10
Value of a: 11
Value of a: 12
```

Extern Storage Class

The extern storage class is used to declare global variables that are accessible from multiple files. When a global variable is declared as extern, it can be used in any other file of the same program by simply declaring it as extern again.

Example:

```
File: extern_variable.c
#include <stdio.h>
int a = 10;
void print_extern_variable()
{
 printf("Value of a is: %d\n", a);
}
File: main.c
#include <stdio.h>
extern int a;
extern void print_extern_variable();
int main()
{
```

```
 print_extern_variable();
 return 0;
}
```

Output:

```
Value of a is: 10
```

6.3 PARAMETER PASSING IN C PROGRAMMING

In C programming, a function is a block of code that performs a specific task and can be called from other parts of a program. When a function is called, values can be passed to it as parameters. These values are referred to as arguments and are used by the function to perform its task.

There are three methods of parameter passing in C programming:

1. Call by value
2. Call by reference
3. Call by pointer

1. Call by value

Call by value is the simplest and most commonly used method of parameter passing in C programming. In this method, a copy of the argument is passed to the function. This means that any changes made to the parameter inside the function will not affect the original argument.

Example:

```
#include <stdio.h>
void swap(int x, int y)
{
 int temp = x;
 x = y;
 y = temp;
}

int main() {
 int a = 5, b = 10;
```

```
    printf("Before swapping: a = %d, b = %d\n", a, b);
    swap(a, b);
    printf("After swapping: a = %d, b = %d\n", a, b);
    return 0;
}
```

Output:

```
Before swapping: a = 5, b = 10
After swapping: a = 5, b = 10
```

2. Call by reference

Call by reference is similar to call by value, but instead of passing a copy of the argument, a reference to the original argument is passed to the function. This means that any changes made to the parameter inside the function will affect the original argument.

Example:

```
#include <stdio.h>
void swap(int *x, int *y)
 {
  int temp = *x;
  *x = *y;
  *y = temp;
}
int main()
{
  int a = 5, b = 10;
  printf("Before swapping: a = %d, b = %d\n", a, b);
  swap(&a, &b);
  printf("After swapping: a = %d, b = %d\n", a, b);
  return 0;
}
```

Output:

```
Before swapping: a = 5, b = 10
After swapping: a = 10, b = 5
```

3. Call by pointer

Call by pointer is similar to call by reference, but instead of passing a reference to the original argument, a pointer to the argument is passed to the function. This means that any changes made to the parameter inside the function will affect the original argument.

Example:

Consider the following code, which swaps the values of two variables:

```
#include <stdio.h>
void swap(int *a, int *b)
{
 int temp = *a;
 *a = *b;
 *b = temp;
}
int main()
{
 int x = 10, y = 20;
 printf("Before swapping: x = %d, y = %d\n", x, y);
 swap(&x, &y);
 printf("After swapping: x = %d, y = %d\n", x, y);
 return 0;
}
```

In this example, the function swap receives the addresses of the x and y variables. The function uses the pointers a and b to access the values of these variables and swaps their values. The result of this code is that the values of x and y are swapped, and the output of the code is:

Before swapping: x = 10, y = 20

After swapping: x = 20, y = 10

6.4 RECURSION

Recursion is a technique in computer programming where a function calls itself in order to solve a problem. This approach of solving problems is quite different from the conventional iterative approach, where a problem is solved by repeating a set of instructions until a certain condition is met. The key idea behind recursion is to divide a complex problem into smaller sub-problems and solving them in a recursive manner.

Why use Recursion?

There are several advantages to using recursion. Firstly, recursion provides a clean and elegant solution to problems that can be difficult to solve iteratively. Secondly, recursion can simplify the code and make it more readable. Finally, some problems are naturally suited for a recursive solution and cannot be easily solved using an iterative approach.

When to use Recursion?

Recursion is particularly useful when solving problems that have a repetitive nature. For example, when finding the factorial of a number, each intermediate result depends on the result of the previous step. Similarly, when traversing a tree structure, each node is processed in a recursive manner, with the solution being built up as the recursion unwinds.

How Recursion Works?

Recursion works by dividing a complex problem into smaller sub-problems and solving them in a recursive manner. The function calls itself until it reaches a base case, which is a condition that stops the recursion. The base case provides the solution to the problem, and the results are combined as the recursion unwinds.

Example:

Factorial of a Number :Let's take an example of finding the factorial of a number to understand how recursion works. The factorial of a number n is given by n! = n * (n-1) * (n-2) * ... * 2 * 1. To find the factorial of a number using recursion, we need to define a function that takes an integer n as an argument and returns n!. The function would then call itself with the argument n-1, until the base case is reached. The base case is defined as n = 0, and the factorial of 0 is 1. The solution is built up as the recursion unwinds.

Here's the code for finding the factorial of a number using recursion in C:

```
#include <stdio.h>
int factorial(int n) {
 if (n == 0) {
 return 1;
 }
 return n * factorial(n-1);
}
int main() {
 int n;
 printf("Enter a number: ");
 scanf("%d", &n);
 printf("The factorial of %d is %d\n", n, factorial(n));
 return 0;
}
```

Explanation: In the above code, the function factorial takes an integer n as an argument and returns the factorial of n. The function first checks if n is equal to 0, and if it is, it returns 1, which is the base case. If n is not equal to 0, the function returns n * factorial(n-1). This is the recursive step, where the function calls itself with the argument n-1. This process continues until the base case is reached, and the solution is built up as the recursion unwinds.

EXERCISE

Short Questions (2 marks)

1. What is a function in programming?
2. What is the difference between user-defined and built-in functions?
3. What is the purpose of storage classes in C++?
4. What is call by value in function parameter passing?
5. What is the idea of call by reference in passing arrays to functions?

Long Questions (5 marks)

1. What is recursion? Explain with an example.
2. What is the difference between automatic, static, and dynamic storage classes? Give an example of each.
3. What is function overloading? How is it useful in programming?
4. What is a default argument in a function? How is it used in programming?
5. What are inline functions? Explain their advantages and disadvantages.

Multiple Choice Questions

1. Which of the following statements is true about user-defined functions?
   a) They are always written in the main() function.
   b) They are created by the compiler.
   c) They are defined by the user.
   d) They are available in all programming languages.

Answer: c) They are defined by the user.

2. Which storage class in C++ is used to declare a variable that retains its value between function calls?
   a) Automatic
   b) Static
   c) Register
   d) External

Answer: b) Static

3. What is call by value in parameter passing?
   a) The actual parameter's address is passed to the function.
   b) A copy of the actual parameter's value is passed to the function.
   c) A reference to the actual parameter is passed to the function.
   d) None of the above.

Answer: b) A copy of the actual parameter's value is passed to the function.

4. What is the difference between call by value and call by reference in parameter passing?
   a) Call by value passes a copy of the actual parameter's value, while call by reference passes the actual parameter's address.
   b) Call by value passes the actual parameter's address, while call by reference passes a copy of the actual parameter's value.
   c) There is no difference between call by value and call by reference.
   d) Call by value is used for built-in functions, while call by reference is used for user-defined functions.

Answer: a) Call by value passes a copy of the actual parameter's value, while call by reference passes the actual parameter's address.

5. What is the output of the following code?

```
void print_array(int arr[], int size) {
 for (int i = 0; i < size; i++) {
 cout << arr[i] << " ";
 }
}
int main() {
 int arr[5] = {1, 2, 3, 4, 5};
 print_array(arr, 5);
 return 0;
}
```

   a) 1 2 3 4 5
   b) 5 4 3 2 1
   c) 1 3 5 7 9
   d) Compiler error

Answer: a) 1 2 3 4 5

6. Which of the following storage classes has the lowest priority?
   a) Static
   b) Register

c) Automatic
d) External

Answer: c) Automatic

7. What is recursion?
    a) A function that returns a value.
    b) A function that calls itself.
    c) A built-in function in C++.
    d) A function that takes no arguments.

Answer: b) A function that calls itself.

7

Strings

7.1 INTRODUCTION

Strings are an essential aspect of any programming language and C programming is no exception. A string in C programming is a sequence of characters stored in a single array, terminated by a null character. In this chapter, we will delve into the details of strings in C programming, including its syntax, operations, and examples.

7.2 DECLARING AND INITIALIZING STRINGS

A string in C programming can be declared as an array of characters. The size of the array must be one more than the number of characters in the string to accommodate the null character that terminates the string. For example, the following code declares and initializes a string "Hello, World!"

```
char greeting[14] = "Hello, World!";
```

Alternatively, you can also declare a string using a pointer to a character, as shown below:

```
char *greeting = "Hello, World!";
```

In this case, the string literal "Hello, World!" is stored in read-only memory, and the pointer greeting points to the first character of the string.

7.3 STRING OPERATIONS

C programming provides several string operations that allow you to manipulate strings in various ways. Some of the commonly used string operations are discussed below.

1. strlen() - This function calculates the length of a string, excluding the null character. For example, the following code calculates the length of the string "Hello, World!"

```
#include <string.h>
#include <stdio.h>
int main()
{
 char greeting[14] = "Hello, World!";
 int length = strlen(greeting);
 printf("The length of the string is: %d\n", length);
 return 0;
}
```

Output:

The length of the string is: 13

2. strcpy() - This function copies one string to another. For example, the following code copies the string "Hello, World!" to the string "Greeting."

```
#include <string.h>
#include <stdio.h>
int main()
{
 char greeting[14] = "Hello, World!";
 char copy[14];
 strcpy(copy, greeting);
 printf("The copied string is: %s\n", copy);
 return 0;
}
```

Output:

The copied string is: Hello, World!

3. strcat() - This function concatenates two strings. For example, the following code concatenates the strings "Hello, " and "World!"

```
#include <string.h>
#include <stdio.h>
int main()
{
 char greeting[20] = "Hello, ";
 strcat(greeting, "World!");
 printf("The concatenated string is: %s\n", greeting);
 return 0;
}
```

Output:

The concatenated string is: Hello, World!

In this chapter, we have covered the basics of strings in C programming, including its syntax, operations, and examples. We have discussed how to declare and initialize strings, as well as how to perform common string operations like calculating the length of a string, copying one string to another, and concatenation operation on string.

EXERCISE

Short questions (2 marks each)

1. What is an array of characters?
2. What is a variable length character string?
3. How do you input a character string?
4. What are character library functions?
5. What are string handling functions?

Long questions (5 marks each)

1. Explain the difference between a fixed length character string and a variable length character string.
2. Write a program to input a character string and count the number of vowels in it.
3. Explain the use of the strcmp() function in C language with an example.
4. What are the different ways of initializing a character string in C language? Explain with examples.
5. Write a program in C language to reverse a given character string using a loop.

Multiple Choice Questions

1. Which of the following is not a string handling function in C language?
    a. strcpy()
    b. strcat()
    c. scanf()
    d. strlen()

Answer: c. scanf()

2. Which function is used to concatenate two strings in C language?
    a. strcat()
    b. strcmp()
    c. strlen()
    d. strcpy()

Answer: a. strcat()

3. What is the output of the following code?

```
char str1[] = "Hello";
char str2[] = "World";
strcat(str1, str2);
printf("%s", str1);
```

    a. HelloWorld
    b. WorldHello

c. elloHWorl

d. The code will not compile.

Answer: a. HelloWorld

4. Which function is used to compare two strings in C language?

    a. strcat()

    b. strcmp()

    c. strlen()

    d. strcpy()

Answer: b. strcmp()

5. What is the output of the following code?

```
char str[] = "hello";
printf("%c", str[1]);
```

    a. h

    b. e

    c. l

    d. o

Answer: b. e

6. Which function is used to find the length of a string in C language?

    a. strcat()

    b. strcmp()

    c. strlen()

    d. strcpy()

Answer: c. strlen()

7. What is the output of the following code?

```
char str[] = "Hello World";
printf("%d", strlen(str));
```

    a. 10

    b. 11

    c. 12

    d. 13

Answer: d. 13

8. Which function is used to copy one string to another in C language?
    a. strcat()
    b. strcmp()
    c. strlen()
    d. strcpy()

Answer: d. strcpy()

9. What is the output of the following code?

```
char str[] = "Hello World";
printf("%s", str + 6);
```

    a. Hello
    b. World
    c. Hello World
    d. The code will not compile.

Answer: b. World

10. What is the output of the following code?

```
char str[] = "Hello";
str[3] = 'p';
printf("%s", str);
```

    a. Hellp
    b. Helop
    c. Helpo
    d. The code will not compile.

Answer: a. Hellp

8

Pointers

8.1 INTRODUCTION

Pointers are a fundamental concept in the C programming language. They allow you to manipulate memory addresses directly, which is often necessary for efficient and flexible programming. In this chapter, we will introduce the basics of pointers and how they are used in C.

8.2 WHAT IS A POINTER?

A pointer is a variable that holds the memory address of another variable. It is declared using the '*' operator, and its value is the memory address of the variable it points to. The following code shows how to declare a pointer in C:

```
int x = 10;
int *ptr;
ptr = &x;
```

In this example, ptr is a pointer to an int and is assigned the memory address of x using the & operator, which returns the memory address of a variable.

Dereferencing a Pointer

The process of accessing the value stored at a memory address is known as dereferencing. This can be done using the '*' operator, as follows:

```
int x = 10;
int *ptr;
ptr = &x;
printf("Value of x: %d\n", *ptr);
```

The output of this code will be:

```
Value of x: 10
```

Pass by Reference

In C, functions can receive arguments either by value or by reference. When a function receives an argument by value, a copy of the argument is passed to the function. On the other hand, when a function receives an argument by reference, the function receives a pointer to the argument, and can modify its value directly. The following code demonstrates how to pass an argument by reference:

```
void swap(int *a, int *b)
{
 int temp = *a;
 *a = *b;
 *b = temp;
}

int main()
{
 int x = 10;
 int y = 20;
 swap(&x, &y);
 printf("x: %d, y: %d\n", x, y);
 return 0;
}
```

The output of this code will be:

```
x: 20, y: 10.
```

Pointer Arithmetic

Pointer arithmetic is a fundamental concept in C programming and is closely related to the manipulation of memory addresses. It involves performing arithmetic operations on pointers, such as incrementing, decrementing, and calculating the difference between two pointers. Understanding pointer arithmetic is crucial for managing memory effectively in C and for writing efficient and effective C programs

Before diving into the rules of pointer arithmetic, it is important to understand the basic syntax of pointers in C. A pointer is a variable that stores the memory address of another variable. The **syntax** for declaring a pointer in C is as follows:

```
data_type *pointer_name;
```

For example, to declare a pointer to an int, the syntax would be:

```
int *ptr;
```

To assign the memory address of a variable to a pointer, you use the address-of operator (&) as follows:

```
int num = 10;
int *ptr = &num;
```

Now that we have covered the basics of pointers, let's move on to the rules of pointer arithmetic.

1. Increment and Decrement: Pointers can be incremented or decremented just like any other variable. When a pointer is incremented, its value is increased by the size of the data type it points to. For example, if a pointer points to an int, its value will be increased by 4 bytes (on most platforms). Similarly, when a pointer is decremented, its value is decreased by the size of the data type it points to.

```
int num = 10;
int *ptr = &num;
ptr++;
```

2. Addition and Subtraction: You can add or subtract an integer value to or from a pointer. This allows you to move the pointer to different memory addresses. The result of the addition or subtraction will be a pointer of the same data type as the original pointer.

```
int num = 10;
int *ptr = &num;
ptr = ptr + 2;
```

3. Comparison: Pointers can be compared using the relational operators (<, >, <=, >=, ==, !=). When two pointers are compared, the result depends on the relative positions of the memory addresses they point to in memory.

```
int num1 = 10;
int num2 = 20;
int *ptr1 = &num1;
int *ptr2 = &num2;
if (ptr1 < ptr2)
{
 printf("ptr1 points to a memory address lower than ptr2");
}
```

4. Subtraction of Pointers: When two pointers of the same data type are subtracted, the result is the number of elements between the two memory addresses they point to, divided by the size of the data type.

```
int nums[10] = {1, 2, 3, 4, 5, 6, 7, 8, 9, 10};
int *ptr1 = &nums[0];
int *ptr2 = &nums[5];
int difference = ptr2 - ptr1;
printf("The number of elements between ptr1 and ptr2 is
%d\n", difference);
```

8.3 APPLICATIONS OF POINTER ARITHMETIC

Pointer arithmetic has many practical applications in real-world programming. Some common uses of pointer arithmetic include:

- **Dynamic memory allocation**: Pointers are often used in conjunction with malloc and free functions to dynamically allocate memory for variables during runtime.
- **Linked lists:** Pointers are a fundamental part of linked lists, where each node in the list contains a pointer to the next node.

- **String manipulation:** Pointers can be used to manipulate strings in C and C++, where strings are represented as arrays of characters.
- **Image processing:** Pointers can be used to manipulate image data stored in arrays, such as accessing and modifying individual pixels in an image

8.4 POINTERS TO POINTERS

Pointers to pointers, also known as double pointers, are variables that store the memory address of other pointers. They are an important concept in C programming, and are useful in dynamic memory allocation and other advanced programming techniques.

Syntax:

Declaring a pointer to pointer in C is done using two asterisks (**). For example:

```
int **ptr;
```

This declares a pointer to pointer of type int, which means it stores the memory address of another pointer of type int.

Accessing the Value of a Pointer to Pointer

To access the value stored in a pointer to pointer, we need to dereference it twice. The first dereference operation gets the value of the pointer stored at the address pointed to by the pointer to pointer, while the second dereference operation gets the value stored at the address pointed to by the first pointer.

The **syntax** for accessing the value of a pointer to pointer is as follows:

```
*(*ptr)
```

Example:

```
#include <stdio.h>
int main ()
{
 int var = 10;
int *ptr;
int **ptr2;
```

```
ptr = &var;
ptr2 = &ptr;
printf("Value of var = %d\n", var);
printf("Value of *ptr = %d\n", *ptr);
printf ("Value of **ptr2 = %d\n", **ptr2);
 return 0;
 }
```

Output:

```
Value of var = 10
Value of *ptr = 10
Value of **ptr2 = 10
```

In the above example, we declare an integer variable **var** and initialize it with the value 10. We then declare a pointer **ptr** to **var** and assign its address to **ptr**. Next, we declare a pointer-to-pointer **ptr2** and assign the address of **ptr** to it.

Finally, we use the printf () function to display the value of **var**, ***ptr**, and ****ptr2**. As we can see from the output, all three values are equal to 10, which confirms that we have correctly dereferenced the pointers to pointers.

Rules for Pointers to Pointers

- A pointer to pointer must be declared with two asterisks (**).
- A pointer to pointer stores the memory address of another pointer.
- To access the value stored in a pointer to pointer, we must dereference it twice.
- The first dereference operation gets the value of the pointer stored at the address pointed to by the pointer to pointer.
- The second dereference operation gets the value stored at the address pointed to by the first pointer.
- Pointers to pointers can be used for dynamic memory allocation, for storing pointers in arrays, and for passing pointers as function arguments.
- When passing a pointer to pointer as a function argument, it must be declared with two asterisks in both the function definition and the function call.

8.5 GENERIC POINTERS

Pointers are a fundamental concept in C and C++ programming languages. They allow you to manipulate memory addresses directly and perform low-level operations on data stored in memory. A pointer variable holds the memory address of another variable of a specific data type. The data type of the variable to which a pointer points is important because it determines the size of memory that should be allocated for that variable and the type of operations that can be performed on it.

However, in some cases, you may want to use a pointer that can point to a variable of any data type. This is where generic pointers come into play. A generic pointer is a type of pointer that can point to a variable of any data type. The most commonly used generic pointer in C is the void pointer.

Here will cover the basics of generic pointers, including what they are, how they work, and when they should be used. We will also explore the void pointer in detail and learn how to cast it to a pointer of a specific data type.

What is a Generic Pointer?

A generic pointer is a type of pointer that can point to a variable of any data type. This makes them extremely flexible and useful in situations where you want to manipulate data without knowing its type. For example, you may want to pass a pointer to a function that should perform some operation on it, but the data type of that pointer is unknown. In this case, you can use a generic pointer to pass the pointer to the function without specifying its data type.

The void Pointer

The void pointer is the most commonly used generic pointer in C. It is a pointer that has no associated data type, and as such, it can point to a variable of any data type.

The void pointer is declared using the void keyword, as follows:

```
void *ptr;
```

When to Use Generic Pointers

Generic pointers should be used in situations where you want to manipulate data without knowing its type. Some of the most common use cases for generic pointers are:

- When you want to pass a pointer to a function as an argument, but the data type of that pointer is unknown.
- When you want to allocate memory dynamically for an unknown data type.
- When you want to store data of different data types in an array or a data structure.

Casting a void Pointer

Since the void pointer has no associated data type, you need to cast it to a pointer of a specific data type before you can use it to access the data it points to.

The **syntax** for casting a void pointer is as follows:

```
(data type *)ptr;
```

For example, if you have a void pointer ptr that you want to cast to a pointer to an int, you would do the following:

```
int *int_ptr = (int *)ptr;
```

In conclusion, generic pointers, and void pointers in particular, are a powerful tool in C and C++ programming languages. They allow you to manipulate data without knowing its type, making them extremely flexible and useful in a variety of situations.

Array of Pointers

In computer programming, an array of pointers is a data structure that consists of an array of variables that are pointers. A pointer is a type of data in computer programming that holds the memory address of a value. An array of pointers allows you to store multiple memory addresses in a single data structure, making it easy to manipulate and manage multiple values at once.

One of the most common uses for an array of pointers is for storing strings. In many programming languages, a string is an array of characters, and an array of pointers can be used to store multiple strings. Each element in the array of pointers holds the memory address of a string, which can be accessed using the element's index. This makes it possible to manipulate multiple strings as a single unit, rather than having to manage each string individually.

For **exampl**e, consider the following C code:

```
#include <stdio.h>
int main()
```

```
{
 char *str_array[3];
 str_array[0] = "First String";
 str_array[1] = "Second String";
 str_array[2] = "Third String";

 for (int i = 0; i < 3; i++) {
 printf ("%s\n", str_array[i]);
 }
 return 0;
}
```

In this code, str_array is an array of pointers that stores three strings. The first element of str_array holds the memory address of the string “First String”, the second element holds the memory address of the string “Second String”, and so on. The for loop is used to access each element of str_array and print the string stored at that memory address.

Another common use for arrays of pointers is for creating dynamic data structures, such as linked lists and trees. In these data structures, each element in the array of pointers holds the memory address of another element in the data structure, creating a chain of pointers that can be used to traverse the data structure.

For example, consider the following code, which creates a linked list in C:

```
#include <stdio.h>
#include <stdlib.h>
struct node
{
 int data;
 struct node *next;
};
int main ()
{
 struct node *head, *tail;
 head = (struct node *) malloc (size of (struct node));
```

```
    tail = (struct node *) malloc (size of (struct node));
    head->data = 1;
    tail->data = 2;
    head->next = tail;
    tail->next = NULL;
    struct node *ptr = head;
    while (ptr != NULL)
   {
    printf("%d\n", ptr->data);
    ptr = ptr->next;
    }

    return 0;
   }
```

In this code, each struct node holds a value (data) and a pointer (next) to the next element in the linked list. The two malloc calls are used to allocate memory for the head and tail of the linked list, and the next pointers are used to link the two elements together. The while loop is used to traverse the linked list and print the data stored in each node.

In conclusion, arrays of pointers are a powerful and flexible data structure in computer programming that can be used for a wide range of purposes, including storing strings and creating dynamic

Function Returning Pointer

In C programming, a function can return a pointer as its result. This is a powerful feature that enables the programmer to manipulate dynamically allocated memory, among other things. A pointer is a variable that stores the memory address of another variable. In C, a function can return a pointer to any data type, including primitive data types and user-defined data types.

Syntax

The syntax for declaring a function that returns a pointer is as follows:

```
<data type> *function_name(parameters)
{
 // code return <pointer to data type>;
}
```

The data type of the pointer returned by the function must match the data type of the pointer declared in the function definition.

Example

Consider the following example of a function that returns a pointer to an int data type:

```
#include <stdio.h>
#include <stdlib.h>
int *create_array(int size)
{
int *array = (int *)malloc(size * sizeof(int));
 return array;
}
int main()
 { int *array;
int size = 5;
array = create_array(size);
 for (int i = 0; i < size; i++)
{
 array[i] = i;
}
for (int i = 0; i < size; i++)
{
 printf("%d ", array[i]);
}
printf("\n");
free(array);
return 0;
}
```

In this example, the **create_array** function returns a pointer to an int data type. The function uses the **malloc** function to dynamically allocate memory for an array of **size** int elements. The **malloc** function returns a pointer to the first byte of the memory block it allocates. The pointer is stored in the **array** variable, which is then returned by the **create_array** function.

In the **main** function, the **create_array** function is called with an argument of **5**, which specifies the size of the array to be created. The result of the function is stored in the **array** variable, which is then used to access the elements of the array. The elements of the array are initialized to the values 0 through 4, and then printed to the console. Finally, the **free** function is called to deallocate the memory block previously allocated by **malloc**.

Rules for Function Returning Pointer

- The pointer returned by the function must be dynamically allocated using **malloc** or **calloc** and must be deallocated using the **free** function after it is no longer needed.
- The pointer returned by the function must not be a local variable declared within the function, as local variables are automatically destroyed when the function returns.
- The pointer returned by the function must not be a static variable, as static variables persist even after the function returns.
- The pointer returned by the function must be assigned to a variable of the same data type as the pointer returned by the function.
- The pointer returned by the function must be used with care, as dereferencing an invalid pointer can result in a segmentation fault.

8.6 DYNAMIC MEMORY ALLOCATION

Dynamic memory allocation is a powerful feature of the C programming language that allows for the allocation of memory at runtime. It enables programs to allocate and deallocate memory as required, instead of having to allocate all memory at the start of the program. This results in more efficient memory utilization and can lead to more flexible and dynamic programs.

Dynamic memory allocation is performed using four library functions: malloc(), calloc(), realloc(), and free(). These functions are part of the C Standard Library and are used to dynamically allocate memory during runtime.

1. malloc() Function

The malloc() function is used to allocate a block of memory with a specified size. The size is specified in bytes and the function returns a pointer to the first byte of the allocated block.

The **syntax** for the malloc() function is as follows:

```
void *malloc(size_t size);
```

Here, **size** specifies the size of the memory block to be allocated, and the function returns a void pointer to the first byte of the allocated memory block.

Example:

```
#include <stdio.h>
 #include <stdlib.h>
 int main()
{
int *ptr;
 int size = 10;
ptr = (int*)malloc(size * sizeof(int));
if(ptr == NULL)
 {
 printf("Memory allocation failed.\n");
 return 0;
 }
printf("Memory allocation successful.\n");
return 0;
}
```

In this example, we allocate memory for an integer array of size 10 using the malloc () function. We first cast the return value of malloc () to an integer pointer, and then check if the pointer is equal to NULL, which indicates a failure in memory allocation. If the allocation is successful, the program prints a message indicating that the allocation was successful.

2. calloc() Function

The calloc() function is used to allocate a block of memory for an array of elements with a specified size, and initializes all elements to 0.

The **syntax** for the calloc() function is as follows:

```
void *calloc(size_t nmemb, size_t size);
```

Here, **nmemb** specifies the number of elements in the array and **size** specifies the size of each element. The function returns a void pointer to the first byte of the allocated memory block.

Example:

```
#include <stdio.h>
#include <stdlib.h>
int main()
{
int *ptr;
int nmemb = 10, size = sizeof(int);
ptr = (int*)calloc(nmemb, size);
if(ptr == NULL)
{
printf("Memory allocation failed.\n");
return 0;
}
printf("Memory allocation successful.\n");
return 0;
}
```

In this example, we allocate memory for an integer array of size 10 using the calloc() function. We first cast the return value of calloc() to an integer pointer, and then check if the pointer is equal to NULL, which indicates a failure in memory allocation. If the allocation is successful, the program prints a message indicating that the allocation was successful.

3. realloc() Function

The realloc() function is used to resize a previously allocated block of memory. The function is defined in the stdlib.h header file. The syntax of the realloc() function is as follows:

```
void *realloc(void *ptr, size_t size);
```

The first argument to realloc() is a pointer to a block of memory previously allocated with malloc(), calloc(), or realloc(). The second argument is the new size in bytes of the block of memory. If the block of memory is successfully resized, realloc() returns a pointer to the resized block of memory. If the memory block cannot be resized, realloc() returns a null pointer.

It's important to note that realloc() may return a different pointer than the one passed as the first argument. This means that after resizing the memory block, the old pointer should no longer be used, and the new pointer returned by realloc() should be used instead.

Example:

```
#include <stdio.h>
 #include <stdlib.h>
int main()
{
int *ptr, i, n;
printf("Enter number of elements: ");
scanf("%d", &n);
 ptr = (int*) malloc(n * sizeof(int));
 if(ptr == NULL)
{
 printf("Error! memory not allocated.");
exit(0);
}
printf("Enter elements of array: ");
 for(i = 0; i < n; ++i)
 {
 scanf("%d", ptr + i);
}
 printf("Array elements before reallocation: ");
for(i = 0; i < n; ++i)
{
 printf("%d ", *(ptr + i));
}
ptr = realloc(ptr, 2 * n * sizeof(int));
```

```
if(ptr == NULL)
{
printf("Error! memory not reallocated.");
 exit (0);
 }
 for (i = n; i < 2 * n; ++i)
 {
 *(ptr + i) = 0;
 }
printf("\nArray elements after reallocation: ");
for (i = 0; i < 2 * n; ++i)
{
printf("%d ", *(ptr + i));
free(ptr);
return 0;
}
```

In this example, we first allocate memory for an array of size n using malloc (). We then prompt the user to enter the elements of the array. Next, we resize the memory block using realloc(), and set the newly allocated memory to zero. Finally, we print the elements of the array both before and after reallocation.

4. free() Function

free function is used to release dynamically allocated memory. Top of Form

The syntax for the **free** function is as follows:

```
void free(void *ptr);
```

Where **ptr** is a pointer to the memory space that needs to be released.

Usage The **free** function is used to release the memory space previously allocated using **malloc** or **calloc**. The **ptr** argument is a pointer to the memory space that needs to be released. It is important to note that the **free** function only releases memory space and does not set its value to zero.

Example Here is an example that demonstrates the use of the **free** function:

```
#include <stdio.h>
#include <stdlib.h>
int main()
{
int n;
int *ptr; printf("Enter the number of elements: ");
scanf("%d", &n);
ptr = (int*) malloc(n * sizeof(int));
if (ptr == NULL)
{
printf("Memory allocation failed\n");
exit(0);
}
 printf("Memory successfully allocated using malloc\n");
 memory free(ptr);
 printf("Memory freed\n");
return 0;
 }
```

In this example, **malloc** is used to allocate memory space for **n** elements of type **int**, and the pointer **ptr** is used to reference this memory space. After using the dynamically allocated memory, the **free** function is called to release the memory space.

When using the **free** function, there are a few important rules to keep in mind:

- The **free** function should only be used on memory spaces previously allocated using **malloc** or **calloc**.
- The same pointer passed to **malloc** or **calloc** should be passed to the **free** function to release the same memory space.
- The **free** function should only be called once for a given memory space. Calling **free** multiple times on the same memory space can lead to undefined behaviour.
- After calling the **free** function, the pointer passed to it should no longer be used.
- The **free** function does not set the value of the memory space to zero, so it is the responsibility of the programmer to set its value to zero if necessary.

- The **free** function is an important part of dynamic memory allocation in the C programming language, and it is essential to understand its syntax, usage, and important rules. By using the **free** function correctly, a program can dynamically allocate memory as required, while also releasing memory space when it is no longer needed, leading to efficient memory management and avoiding memory leaks.

8.7 POINTERS TO FUNCTIONS

Pointers are an essential part of C programming language, and when combined with functions, they offer even more versatility and power. Pointers to functions, also known as function pointers, are pointers that hold addresses of functions. These function pointers can be used to call functions dynamically, or store a set of functions to call at a later time.

Function pointers in C programming offer a flexible and powerful way to work with functions. They provide the ability to call functions dynamically, store a set of functions to call at a later time, and pass functions as arguments to other functions. Function pointers are used extensively in various programming applications, including event-driven programming, dynamic libraries, and more.

When using function pointers, it is important to ensure that the function prototype of the function pointer matches the function it is pointing to. Incorrect function prototypes can result in unexpected behaviour and errors during runtime.

The syntax of declaring a function pointer is similar to declaring a normal pointer, with the difference being the addition of the function prototype within parentheses after the * operator. The function prototype defines the return type and the number and type of arguments the function takes.

ReturnType (*pointerName)(ArgumentTypes);

For example, the following code declares a function pointer named "ptr" that points to a function that takes two integers as arguments and returns an integer:

```
int (*ptr)(int, int);
```

Assigning a Function to a Function Pointer: A function pointer can be assigned to a function of the same prototype. For example, the following code assigns the add function to the function pointer "ptr":

```
int add(int a, int b)
```

```
{
return a + b;
}
int (*ptr)(int, int) = add;
```

Using a Function Pointer to Call a Function: A function pointer can be used to call a function just like a normal function call. For example, the following code calls the add function through the function pointer "ptr":

```
int result = (*ptr)(5, 10);
```

Passing a Function Pointer as an Argument: Function pointers can also be passed as arguments to other functions. For example, the following code declares a function that takes a function pointer as an argument and calls it:

```
void call Function(int (*function)(int, int), int a, int b)
int result = (*function) (a, b);
printf("Result: %d\n", result);
}
```

The following code passes the add function as an argument to the callFunction:

```
int add(int a, int b)
{
return a + b;
}
int (*ptr)(int, int) = add;
callFunction(ptr, 5, 10);
```

Array of Function Pointers: An array of function pointers can be used to store a set of functions to call later.

For example, the following code declares an array of function pointers named "functions":

```
int (*functions[3])(int, int);
```

The following code assigns three functions to the array:

```
int add(int a, int b)
{
return a + b;
}
```

```
int subtract(int a, int b)
{
return a - b;
}
int multiply(int a, int b)
{
return a * b;
}
int (*functions[3])(int, int) = {add, subtract, multiply};
```

The following code calls the functions stored in the array:

```
for (int i = 0; i < 3; i++)
{
int result = (*functions[i])(5, 10);
printf("Result: %d\n", result);
}
```

In conclusion, understanding pointers to functions is an essential part of C programming and is a valuable tool for any programmer. With the knowledge of function pointers, one can build more dynamic and flexible applications.

EXERCISE

Short Questions (2 marks each)

1. What is a pointer?
2. How do you declare a pointer variable?
3. What is the size of a pointer variable?
4. What is pointer arithmetic?
5. What is the difference between a void pointer and a NULL pointer?

Long Questions (5 marks each)

1. Explain how pointers are used in C programming with examples.

2. What is pointer arithmetic? Explain with examples.
3. What are pointers to pointers? How are they useful? Give an example.
4. What are generic pointers? Explain with an example.
5. What is an array of pointers? Explain with an example.

Multiple Choice Questions:

1. What is the output of the following code?

```
int *p;
printf("%d", p);
```

   a) 0
   b) Garbage value
   c) Error
   d) None of the above

Answer: b) Garbage value

2. What is the output of the following code?

```
int arr[] = {1, 2, 3, 4, 5};
int *p = arr;
printf("%d", *(p+2));
```

   a) 1
   b) 2
   c) 3
   d) 4

Answer: c) 3

3. Which of the following statements is true about pointers?
   a) A pointer can point to multiple variables at the same time.
   b) A pointer can point to a variable of a different data type.
   c) A pointer can be assigned to an integer value.
   d) A pointer cannot be assigned to a NULL value.

Answer: b) A pointer can point to a variable of a different data type.

4. 4.What is the output of the following code?

```
int *p;
int **q;
```

```
q = &p;
printf("%d", **q);
```

a) 0
b) Garbage value
c) Error
d) None of the above

Answer: b) Garbage value

5. What is the output of the following code?

```
int arr[] = {1, 2, 3, 4, 5};
int *p[] = {&arr[0], &arr[1], &arr[2], &arr[3], &arr[4]};
printf("%d", *p[3]);
```

a) 1
b) 2
c) 3
d) 4

Answer: d) 4

6. What is the output of the following code?

```
int *p;
p = (int*)malloc(sizeof(int));
*p = 10;
printf("%d", *p);
```

a) 0
b) Garbage value
c) 10
d) Error

Answer: c) 10

7. Which of the following statements is true about pointers to functions?
   a) A pointer to a function cannot be assigned to another function.
   b) A pointer to a function can be used to pass a function as an argument to another function.

c) A pointer to a function can only point to functions with the same return type.

d) A pointer to a function can only be used in void functions.

Answer: b) A pointer to a function can be used to pass a function as an argument to another function.

8. What is the output of the following code?

```
char str[] = "hello";
char *p = str;
printf("%s", p);
```

a) hello

b) h

c) e

d) l

Answer: a) hello

9

Structure and Union

9.1 INTRODUCTION

A structure in C is a user-defined data type that can hold different data types within a single unit. A structure is used to group data of different data types under a single name. The structure creates a new data type, which can be used in place of basic data types like int, float, and char, etc. The structure in C allows us to organize our data efficiently and provides us with a way to handle complex data.

The **syntax of the structure in** C is as follows:

```
struct structure_name
{
  data_type member1; data_type member2; . . . data_type
member_n;
};
```

The 'struct' keyword is used to define a structure. The structure_ name is the name of the structure. Within the curly braces, we declare the data members, which are the variables that hold the data. Each member is preceded by its data type and a semicolon.

Declaring Structure Variables

Once we have defined a structure, we can declare variables of that structure type. The syntax for declaring a structure variable is as follows:

```
struct structure_name variable_name;
```

Accessing Structure Members

We can access the members of a structure using the dot operator (.). The dot operator is used to access the members of the structure. The syntax for accessing the members of a structure is as follows:

```
variable_name.member_name;
```

Example of Structure

```
#include <stdio.h>
 #include <string.h>
struct student
{
int roll_no;
char name[20];
int age;
};
int main()
{
struct student s1;
 s1.roll_no = 101;
strcpy(s1.name, "John");
s1.age = 20;
printf("Roll No: %d\n", s1.roll_no);
printf("Name: %s\n", s1.name);
printf("Age: %d\n", s1.age);
 return 0;
}
```

Output:

```
Roll No: 101
Name: John
Age: 20
```

In this example, we have defined a structure named ‘student’ that has three members: roll_no, name, and age. We have declared a structure variable ‘s1’ of the structure type ‘student’. We have assigned values to the members of the structure using the dot operator. Finally, we have printed the values of the members using the printf statement.

9.2 TYPES OF STRUCTURES

There are two types of structures in C programming:

1. Simple Structure: A simple structure is a structure that only contains variables. The syntax for defining a simple structure is as follows:

```
struct structure_name
{
data_type variable_1;
data_type variable_2; ...
};
```

Example:

```
struct student
{
int roll_no;
char name[20];
 float marks;
};
int main()
{
struct student s1;
s1.roll_no = 1
strcpy(s1.name, "John");
s1.marks = 90.5;
printf("Roll No: %d\n", s1.roll_no);
printf("Name: %s\n", s1.name);
printf("Marks: %f\n", s1.marks);
return 0;
 }
```

Output:

```
Roll No: 1
Name: John
Marks: 90.500000
```

2. Nested Structure: A nested structure is a structure that contains another structure as its member. The syntax for defining a nested structure is as follows:

```
struct outer_structure_name
{
data_type variable_1;
struct inner_structure_name
{
data_type variable_2;
data_type variable_3;
}
 inner_structure_variable;
};
```

Example:

```
struct address
{
 char city[20];
 int pin_code;
};
struct student
{
int roll_no;
char name[20];
 struct address addr;
};
int main()
{
```

```
 struct student s1;
 s1.roll_no = 1;
strcpy(s1.name, "John");
strcpy(s1.addr.city, "New York");
s1.addr.pin_code = 123456;
printf("Roll No: %d\n", s1.roll_no);
printf("Name: %s\n", s1.name);
printf("City: %s\n", s1.addr.city);
printf("Pin Code: %d\n", s1.addr.pin_code);
return 0;
}
```

Output:

```
Roll No: 1
Name: John
City: New York
Pin Code: 123456
```

Array of Structures

An array of structures is an array that contains multiple structures as its elements. This concept is useful when we want to store and manage multiple related pieces of information.

The syntax for defining an array of structures is as follows:

```
struct structure_name
{
data_type variable_1;
data_type variable_2; ...
} structure_array_name[array_size];
```

Here, **structure_name** is the name of the structure, **variable_1**, **variable_2**, etc. are the variables inside the structure, and **structure_array_name** is the name of the array of structures. The **array_size** is the number of elements in the array.

Example:

```
struct student
{
```

```
int roll_no;
 char name[20];
 float marks;
};
 int main()
 {
 struct student students[3];
students[0].roll_no = 1;
strcpy(students[0].name, "John");
students[0].marks = 90.5;
 students[1].roll_no = 2;
strcpy(students[1].name, "Jane");
students[1].marks = 92.0;
students[2].roll_no = 3;
strcpy(students[2].name, "Jim");
students[2].marks = 85.0;
for (int i = 0; i < 3; i++)
{
printf("Roll No: %d\n", students[i].roll_no);
printf("Name: %s\n", students[i].name);
printf("Marks: %f\n\n", students[i].marks);
}
 return 0;
 }
```

Output:

```
Roll No: 1 Name: John Marks: 90.500000
Roll No: 2 Name: Jane Marks: 92.000000
Roll No: 3 Name: Jim Marks: 85.000000
```

In the above example, we have defined an array of structures named **students** with 3 elements. Each element of the array is a structure of type **student**. We have assigned values to the variables inside each structure, and then displayed the values using a for loop.

Accessing Elements of an Array of Structures

An array of structures is an array that contains multiple structures as its elements. To access the elements of an array of structures, we use the array index. The syntax for accessing the elements of an array of structures is as follows:

```
structure_array_name[index].member_name;
```

Where **structure_array_name** is the name of the array, **index** is the index of the element, and **member_name** is the name of the structure member.

Example:

```
#include <stdio.h>
struct student
{
int roll_no;
char name[20];
float marks;
};
int main()
{
struct student students[3] = { {1, "John", 90.5}, {2, "Jane", 92.0}, {3, "Jim", 85.0} };
for (int i = 0; i < 3; i++)
{
printf("Details of student %d:\n", i+1);
printf("Roll No: %d\n", students[i].roll_no);
printf("Name: %s\n", students[i].name);
printf("Marks: %.2f\n", students[i].marks);
}
return 0;
}
```

In this example, we have an array of structures named **students** with three elements. We use a **for** loop to iterate through each element of the array and access its members using the **structure_array_name[index].member_name** syntax.

It's worth noting that arrays in C programming are zero-indexed, meaning the first element has an index of 0, the second element has an index of 1, and so on. Therefore, when accessing the elements of an array of structures, we must use an index value between 0 and **array_size - 1**.

In conclusion, accessing the elements of an array of structures is a crucial aspect of working with this data structure in C programming. It allows us to extract the values of the structure members and use them in our program.

Modifying Elements of an Array of Structures

Once we have accessed an element of an array of structures, we can modify its values just like we would with a regular variable. The syntax for modifying the values of a structure member is as follows:

```
structure_array_name[index].member_name = new_value;
```

Where **structure_array_name** is the name of the array, **index** is the index of the element, **member_name** is the name of the structure member, and **new_value** is the value that we want to assign to the structure member.

Example:

```
#include <stdio.h>
 struct student
{
int roll_no;
char name[20];
float marks;
};
 int main()
{
  struct student students[3] = { {1, "John", 90.5}, {2,
"Jane", 92.0}, {3, "Jim", 85.0}
}; // Modifying the marks of the second student
 students[1].marks = 95.0;
for (int i = 0; i < 3; i++)
{
printf("Details of student %d:\n", i+1);
```

```
printf("Roll No: %d\n", students[i].roll_no);
printf("Name: %s\n", students[i].name);
printf("Marks: %.2f\n", students[i].marks);
}
return 0;
}
```

In this example, we have an array of structures named **students** with three elements. Before printing the details of each student, we modify the marks of the second student by assigning a new value to **students[1].marks**. Then, we use a **for** loop to print the details of each student, including the modified marks of the second student.

Modifying the elements of an array of structures is an important aspect of working with this data structure in C programming. It allows us to update the values of the structure members and use the updated values in our program.

9.3 STRUCTURE AND FUNCTION IN C PROGRAMMING

In C programming, structure and function are two essential concepts that are closely related to each other. Structure is a composite data type that allows you to store multiple values of different data types under a single name. Function, on the other hand, is a block of code that performs a specific task and can be called multiple times from various parts of the program. Here we will explore how structure and function are related in C programming and how they can be used together to achieve a desired outcome.

Structures in C

A structure in C is a composite data type that can store multiple values of different data types under a single name. The values stored in a structure are known as members. A structure is defined using the "struct" keyword followed by the structure name and a list of members enclosed in curly braces. For example, the following code defines a structure called "student" with three members: name, age, and grade.

```
struct student
{
char name[30];
int age;
```

```
char grade;
};
```

Once a structure is defined, you can declare variables of that type. For example, the following code declares two variables of type "student".

```
struct student s1, s2;
```

To access the members of a structure, you use the dot operator (.) followed by the member name. For example, the following code sets the value of the "name" member of the "s1" structure.

```
strcpy(s1.name, "John Doe");
```

Functions in C

Functions are blocks of code that perform a specific task. Functions in C can take arguments and return a value. Functions allow you to write modular code that can be reused in multiple parts of your program. A function is defined using the "int" or "void" keyword followed by the function name and a list of arguments enclosed in parentheses. For example, the following code defines a function called "add" that takes two arguments and returns their sum.

```
int add(int a, int b)
{
 int c = a + b;
 return c;
}
```

To call a function, you simply use the function name followed by its arguments enclosed in parentheses. For example, the following code calls the "add" function with the arguments 2 and 3 and stores the result in the variable "sum".

```
int sum = add(2, 3);
```

Passing Structures as Arguments to Functions

One way to use structures and functions together is by passing a structure as an argument to a function. This allows the function to operate on the data stored in the structure. For example, consider the following code that defines a structure called "student" with three members: name, age, and grade.

```
struct student
{
```

```
char name[30];
 int age;
char grade;
};
```

Next, let's define a function called "print_student" that takes a structure of type "student" as an argument and prints its members.

```
void print_student(struct student s)
{
printf("Name: %s\n", s.name);
printf("Age: %d\n", s.age);
 printf("Grade: %c\n", s.grade);
}
```

In the main function, we can declare a variable of type "student", fill it with data, and then pass it as an argument to the "print_student" function.

```
int main()
{
 struct student s1;
strcpy(s1.name, "John Doe");
 s1.age = 25; s1.grade = 'A';
print_student(s1);
 return 0;
}
```

When this code is run, it will print the following output:

```
Name: John Doe Age: 25 Grade: A
```

Returning Structures from Functions

Another way to use structures and functions together is by returning a structure from a function. This allows you to create a new structure in a function and return it to the calling function. For example, consider the following code that defines a function called "create_student" that takes two arguments and returns a structure of type "student".

```
struct student create_student(char *name, int age)
{
```

```
    struct student s;
    strcpy(s.name, name);
    s.age = age; s.grade = 'A';
   return s;
    }
```

In the main function, we can call the "create_student" function and store the returned structure in a variable.

```
   int main()
   {
   struct student s1 = create_student("John Doe", 25);
    print_student(s1);
    return 0;
    }
```

When this code is run, it will print the same output as the previous example.

In conclusion, structure and function are two important concepts in C programming that can be used together to enhance the functionality and organization of your code. By passing structures as arguments to functions and returning structures from functions, you can write modular, reusable code that can be easily maintained and modified. By using these concepts effectively, you can write more efficient and organized code that meets your needs.

9.4 SELF-REFERENTIAL STRUCTURES

Self-referential structures are an important concept in computer programming and they form the backbone of many complex data structures like linked lists, trees, and graphs. A self-referential structure is a structure that contains a pointer to a structure of the same type. This pointer allows the structure to reference itself, thereby forming a chain of structures, each linked to another.

- **Definition:** A self-referential structure is a structure that contains a pointer to a structure of the same type. This pointer allows the structure to reference itself, thereby forming a chain of structures, each linked to another.
- **Declaring a Self-Referential Structure**: To declare a self-referential structure, you need to define a structure type and

then declare a pointer of the same type within the structure. This pointer is then used to reference the next element in the chain.

Example:

```
struct node
{
 int data;
 struct node *next;
};
```

In this example, the structure node contains an integer data and a pointer next of type struct node. The next pointer is used to reference the next element in the chain.

Applications of Self-Referential Structures

Self-referential structures are used to build complex data structures like linked lists, trees, and graphs. A linked list is a linear collection of nodes where each node contains a value and a pointer to the next node in the list. Linked lists are often used to implement dynamic data structures like stacks and queues.

Example:

```
struct node *head;
struct node *temp;
 temp = (struct node *)malloc(sizeof(struct node));
 temp->data = 10;
 temp->next = NULL;
head = temp;
 temp = (struct node *)malloc(sizeof(struct node));
temp->data = 20;
temp->next = head;
 head = temp;
```

In this example, two nodes are created using the malloc function and their data values are set to 10 and 20 respectively. The second node is then linked to the first node by setting its next pointer to head, making the second node the new head of the list.

Advantages of Self-Referential Structures:

Self-referential structures provide a flexible and efficient way to build complex data structures like linked lists, trees, and graphs. These structures allow for dynamic allocation of memory, making them ideal for use in a variety of applications where data needs to be stored and manipulated in an efficient manner.

Self-referential structures are a fundamental concept in computer programming and are an essential tool for building complex data structures like linked lists, trees, and graphs. Understanding how self-referential structures work and how to use them effectively is an important aspect of becoming a skilled programmer. With the knowledge of self-referential structures, you will be able to build dynamic data structures that can be used to solve a wide range of computational problems.

9.5 UNION

In C programming, a union is a user-defined data type that allows you to store multiple values of different data types in the same memory location. The concept of union is useful when you want to save memory space in your program. Unlike structures, where each member has its own memory location, all members of a union share the same memory location. This means that the size of a union is equal to the size of its largest member.

The union keyword is used to declare a union data type, and the members of a union are defined within curly braces {}. The union data type can be used in a similar way as a structure or an array.

Some important points about Union:

- A union can have multiple members, but only one member can be accessed at a time.
- The size of a union is equal to the size of its largest member.
- When you assign a value to a union member, it overwrites the previous value of the union.
- When you declare a union, you must initialize at least one of its members.
- Unions can be used to store different types of data, including character arrays, integer values, and floating-point values.
- Unions can also be used to store bit fields, which are individual bits within a memory location.

Syntax:

The syntax for declaring a union in C programming is as follows:

union union_name { data_type member_name; data_type member_name; ... };

Here, **union_name** is the name of the union data type, and **member_name** is the name of a member of the union. **data_type** is the data type of the member, which can be any valid C data type, including int, float, char, or any user-defined data type.

Here is an example of a union in C programming:

```
#include <stdio.h>
 union example { int x; float y; char z; };
int main() {
union example var;
 var.x = 10;
 printf("The value of var.x is: %d\n", var.x);
 var.y = 10.5;
printf("The value of var.y is: %f\n", var.y);
var.z = 'A';
printf("The value of var.z is: %c\n", var.z);
return 0;
}
```

In this example, we declare a union named **example** with three members, **x**, **y**, and **z**, of type int, float, and char, respectively. The value of **var.x** is assigned the value 10, which is printed using the **printf** function. Similarly, the value of **var.y** is assigned the value 10.5, and the value of **var.z** is assigned the value 'A'.

Output:

The value of var.x is: 10 The value of var.y is: 10.500000 The value of var.z is: A

9.6 STRUCTURES vs UNIONS

The differences between structures and unions in C programming can be summarized in the following table:

Structures	Unions
Each member of a structure has its own memory location.	All members of a union share the same memory location.
The size of a structure is equal to the sum of the sizes of its members.	The size of a union is equal to the size of its largest member.
Members of a structure can be accessed simultaneously.	Only one member of a union can be accessed at a time.
Members of a structure are stored in separate memory locations.	Members of a union are stored in the same memory location.

Typedef

typedef is a keyword in C that allows a programmer to create a new data type identifier based on an existing data type. This new data type identifier can then be used in place of the original data type, making code more readable and reducing the need to repeatedly specify complex data types.

The **syntax** for declaring a typedef is as follows:

```
typedef existing_data_type new_data_type_identifier;
```

For **example**, the following code creates a new data type identifier int_ptr based on the int data type:

```
typedef int *int_ptr;
```

Now, instead of declaring variables as int *, they can be declared as int_ptr, like this:

```
int_ptr x;
```

typedef can also be used with complex data types, such as structures and unions, making code more readable and reducing the need to repeatedly specify complex data types. For example, the following code creates a new data type identifier value based on a union of an int and a float:

```
typedef union {
  int i;
  float f;
} value;
```

Now, instead of declaring variables as union { int i; float f; }, they can be declared as value, like this:

```
value v;
```

In conclusion, typedef provides a way for programmers to create new data type identifiers based on existing data types, making code more readable and reducing the need to repeatedly specify complex data types. It is a valuable tool for improving code clarity and reducing the amount of repetitive code in large projects.

Enumeration

Enumeration is a user-defined data type in C that allows a programmer to assign a set of meaningful names to a set of integer values. The names assigned to the integer values are called enumerators, and each enumerator is assigned a unique integer value, which is called the enumeration constant.

The **syntax** for declaring an enumeration in C is as follows:

```
enum enumeration_name { enumerator_name_1, enumerator_name_2, . . enumerator_name_n };
```

For example, the following enumeration declaration creates an enumeration color with three enumerators: red, green, and blue:

```
enum color { red, green, blue };
```

Once an enumeration is declared, variables can be declared as instances of the enumeration, like this:

```
enum color c;
```

By default, the first enumerator is assigned the value 0, the second enumerator is assigned the value 1, and so on. However, the values assigned to the enumerators can be explicitly specified, like this:

```
enum color { red = 1, green = 2, blue = 4 };
```

Enumerations can be used in operations just like any other data type. For example, the following code assigns the value red to the variable c:

```
c = red;
```

Enumerations can also be used in conditional statements, like this:

```
if (c == red)
{ . . .
}
```

Using enumerations in C can help improve code readability and reduce the need to use numeric constants in code. They can also be used to define a set of values that are used repeatedly throughout a program. For example, in a program that deals with the manipulation of images, an enumeration can be used to define the different image file formats that the program can work with.

It is also important to note that when using enumerations in C, it is good practice to explicitly specify the type of the enumeration variables. This is done using the enum keyword, as shown in the examples above. This helps to prevent unexpected behavior in code due to type mismatches.

In conclusion, enumerations in C provide a useful mechanism for assigning a set of meaningful names to a set of integer values. They help improve code readability and reduce the need to use numeric constants in code. They can be used to define a set of values that are used repeatedly throughout a program, and their use can help improve the maintainability of code. By following best practices, such as explicitly specifying the type of enumeration variables, developers can ensure that their code is robust and reliable.

EXERCISE

Short Questions

1. What is a structure in C programming?
2. How can you access a structure member in C?
3. What is the keyword used to declare a union in C?
4. What is the purpose of typedef in C?
5. What is an enumeration in C programming?

Long Questions

1. What is the difference between a structure and a union in C programming? Provide an example of when you would use a union instead of a structure.
2. How can you initialize a structure in C programming? Provide an example.
3. How can you pass a structure to a function in C programming? Provide an example.
4. How can you create a self-referential structure in C programming? Provide an example and explain its purpose.

5. What are nested structures in C programming? Provide an example and explain how to access a member of a nested structure.

Multiple Choice Questions:

1. What is a structure in C programming?
   a) A data type that can store multiple values of different types.
   b) A data type that can store only one value.
   c) A keyword used to define a function.
   d) None of the above.

Answer: A

2. What is the keyword used to declare a union in C?
   a) union
   b) struct
   c) typedef
   d) enum

Answer: A

3. How can you access a structure member in C?
   a) Using the dot operator (.)
   b) Using the arrow operator (->)
   c) Using the pound operator (#)
   d) Using the star operator (*)

Answer: A

4. What is an enumeration in C programming?
   a) A data type used to represent a set of named integer constants.
   b) A data type used to represent a set of named float constants.
   c) A data type used to represent a set of named string constants.
   d) None of the above.

Answer: A

5. What is the purpose of typedef in C?
   a) To create an alias for an existing data type.
   b) To declare a new data type.

c) To create a constant variable.
d) To declare a function.

Answer: A

6. What is the difference between a structure and a union in C programming?
   a) A structure can store multiple values of different types, while a union can store only one value.
   b) A union can store multiple values of different types, while a structure can store only one value.
   c) There is no difference between a structure and a union.
   d) A structure and a union cannot be used in the same program.

Answer: A

7. How can you initialize a structure in C programming?
   a) Using the equal (=) operator.
   b) Using the asterisk (*) operator.
   c) Using the ampersand (&) operator.
   d) Using the dot (.) operator.

Answer: A

8. How can you pass a structure to a function in C programming?
   a) By passing the structure variable directly.
   b) By passing the structure variable with the ampersand (&) operator.
   c) By passing the structure variable with the asterisk (*) operator.
   d) By passing the structure variable with the dot (.) operator.

Answer: A

9. What are nested structures in C programming?
   a) Structures that are declared inside other structures.
   b) Structures that have the same name as another structure.
   c) Structures that are declared outside of a function.
   d) None of the above.

Answer: A

10

File Handling

10.1 INTRODUCTION

File handling is a crucial aspect of any programming language and it allows a programmer to store data in an organized and structured manner. C programming language offers several inbuilt functions to handle and manipulate files. In this chapter, we will discuss the basic concepts of file handling, different file operations and ways to perform them using C programming.

10.2 FILE MODES IN C

When we perform file operations in C, we must specify the mode in which we want to open the file. The mode determines the type of access that we want to have on the file. There are several file modes available in C programming and some of the most commonly used file modes are:

"r" - Read mode, opens the file for reading.

"w" - Write mode, opens the file for writing. If the file exists, its contents are truncated. If the file does not exist, it is created.

"a" - Append mode, opens the file for writing. If the file exists, data is appended to the end of the file. If the file does not exist, it is created.

"r+" - Read and Write mode, opens the file for reading and writing.

"w+" - Write and Read mode, opens the file for writing and reading. If the file exists, its contents are truncated. If the file does not exist, it is created.

"a+" - Append and Read mode, opens the file for writing and reading. If the file exists, data is appended to the end of the file. If the file does not exist, it is created.

10.3 FILE OPERATIONS IN C

There are several file operations that can be performed using C programming. Some of the most commonly used file operations are:

- Creating a file
- Opening a file
- Reading from a file
- Writing to a file
- Appending to a file
- Closing a file

Each operation is discussed below in detail: -

1. Creating a file

To create a file in C, we can use the fopen() function. The fopen() function takes two parameters - the name of the file and the mode in which the file should be opened. If the file does not exist, it is created. The mode for creating a file is typically "w".

Syntax:

```
FILE *fp;
fp = fopen("filename.txt", "w");
```

```
Example:
#include <stdio.h>
int main()
{
FILE *fp;
fp = fopen("sample.txt", "w");
 if (fp == NULL)
{
```

```
printf("Error: Unable to create file\n");
 return 1;
}
printf("File created successfully\n");
fclose(fp);
return 0;
}
```

In the above example, the file pointer **fp** is used to access the file "sample.txt". It acts as a handle to the file and is used in all file operations that we perform on the file. If the file is successfully created, the fopen() function returns a pointer to the file and we can store it in a file pointer. If the file cannot be created, the fopen() function returns a **NULL** pointer and we can handle the error appropriately.

2. Opening a file

To open an existing file in C, we can use the fopen() function. The fopen() function takes two parameters - the name of the file and the mode in which the file should be opened. The mode for opening a file can be "r" for reading, "w" for writing, "a" for appending, "r+" for reading and writing, "w+" for writing and reading, or "a+" for appending and reading.

Syntax:

```
FILE *fp;
fp = fopen("filename.txt", "mode");
```

Example:

```
#include <stdio.h>
int main()
{
FILE *fp;
fp = fopen("sample.txt", "r");
if (fp == NULL)
 {
printf("Error: Unable to open file\n");
return 1;
}
```

```
printf("File opened successfully\n");
fclose(fp);
return 0;
}
```

In the above example, the file pointer **fp** is used to access the file "sample.txt". It acts as a handle to the file and is used in all file operations that we perform on the file. If the file is successfully opened, the fopen() function returns a pointer to the file and we can store it in a file pointer. If the file cannot be opened, the fopen() function returns a **NULL** pointer and we can handle the error appropriately.

3. Reading from a file

To read data from a file in C, we can use the fscanf() function. The fscanf() function takes two parameters - a pointer to the file and the format specifier for the data that is to be read.

Syntax:

```
fscanf(fp, "format specifier", &variable);
```

Example:

```
#include <stdio.h>
 int main()
{
FILE *fp;
int num;
fp = fopen("sample.txt", "r");
 if (fp == NULL)
 {
printf("Error: Unable to open file\n");
 return 1;
 }
fscanf(fp, "%d", &num);
fprintf("Number read from file: %d\n", num);
 fclose(fp);
return 0;
}
```

In the above example, the file pointer **fp** is used to access the file "sample.txt". It acts as a handle to the file and is used in the fscanf() function to specify the file from which data is to be read. The data is read from the file and stored in the variable **num**.

4. Writing to a file

To write data to a file in C, we can use the fprintf() function. The fprintf() function takes two parameters - a pointer to the file and the format specifier for the data that is to be written.

Syntax:

```
fprintf(fp, "format specifier", variable);
```

Example:

```
#include <stdio.h>
int main()
{
FILE *fp;
int num = 123;
fp = fopen("sample.txt", "w");
if (fp == NULL)
{
printf("Error: Unable to open file\n");
 return 1;
}
fprintf(fp, "%d", num);
printf("Number written to file: %d\n", num);
fclose(fp);
 return 0;
 }
```

In the above example, the file pointer **fp** is used to access the file "sample.txt". It acts as a handle to the file and is used in the fprintf() function to specify the file to which data is to be written. The data stored in the variable **num** is written to the file.

5. Append operation

To append data to a file in C, we can use the fprintf() function with the "a" mode. The "a" mode opens the file for writing and appends the data to the end of the file, preserving any existing data in the file.

Syntax:

```
fp = fopen("file.txt", "a");
fprintf(fp, "format specifier", variable);
```

Example:

```
#include <stdio.h>
 int main()
{
FILE *fp;
int num = 123;
fp = fopen("sample.txt", "a");
if (fp == NULL)
{
printf("Error: Unable to open file\n");
return 1;
 }
 fprintf(fp, "%d\n", num);
printf("Number appended to file: %d\n", num);
fclose(fp);
return 0;
}
```

In the above example, the file pointer **fp** is used to access the file "sample.txt". It acts as a handle to the file and is used in the fprintf() function to specify the file to which data is to be appended. The data stored in the variable **num** is appended to the file.

6. Closing a file

To close a file in C, we can use the fclose() function. The fclose() function takes one parameter - a pointer to the file.

Syntax:

```
fclose(fp);
```

Example:

```
#include <stdio.h>
int main()
{
```

```
FILE *fp;
fp = fopen("sample.txt", "a");
if (fp == NULL)
{
printf("Error: Unable to open file\n");
return 1;
}
fclose(fp);
printf("File closed successfully\n");
return 0;
}
```

In the above example, the file pointer **fp** is used to access the file "sample.txt". It acts as a handle to the file and is used in the fclose() function to specify the file that is to be closed. The file is closed using the fclose() function, releasing any resources that were used by the file.

Table Shows Difference between all the operation

Operation	**Description**	**Syntax**	**Example**
Open	Opens an existing file or creates a new file for reading or writing.	fp = fopen("file.txt", "mode");	fp = fopen("sample. txt", "r");
Read	Reads data from a file.	fscanf(fp, "format specifier", &variable);	fscanf(fp, "%d", &num);
Write	Writes data to a file, replacing any existing data.	fprintf(fp, "format specifier", variable);	fprintf(fp, "%d\n", num);
Append	Writes data to a file, preserving any existing data.	fp = fopen("file.txt", "a");	fp = fopen("sample. txt", "a");
Close	Closes a file, releasing any resources that were used by the file.	fclose(fp);	fclose(fp);

10.4 COMMAND LINE ARGUMENTS

Command line arguments are parameters passed to a program when it is executed from the command line. In C programming, these arguments can be processed and used within the program to modify its behaviour.

Here's how you can use command line arguments in C programming for file handling:

Declare the main function with two parameters: **int main(int argc, char *argv[])**

- **argc** (argument count) is an integer that represents the number of arguments passed to the program, including the name of the program itself.
- **argv** (argument vector) is an array of strings that contains the actual arguments passed to the program. **argv[0]** is the name of the program, **argv[1]** is the first argument, and so on.

Access the arguments within the program using the **argv** array.

For example, if you pass the file name as a command line argument to the program, you can access it using **argv[1]**.

Here is an example of how to use command line arguments in a program that reads data from a file:

```
#include <stdio.h>
int main(int argc, char *argv[])
 {
FILE *fp;
int num;
 if (argc != 2)
{
 printf("Usage: %s filename\n", argv[0]);
 return 1;
}
 fp = fopen(argv[1], "r");
 if (fp == NULL)
{
printf("Could not open file %s\n", argv[1]);
return 1;
}
 while (fscanf(fp, "%d", &num) != EOF)
{
```

```
    printf("%d\n", num);
  }
  fclose(fp);
  return 0;
}
```

In this example, the program checks if the number of arguments passed to it is equal to 2 (the program name and the file name). If not, it prints an error message and exits. If the number of arguments is correct, the program opens the file specified in **argv[1]** and reads its contents, printing each number on a separate line.Top of FormBottom of Form

10.5 SCOPE AND LIFE OF A VARIABLE IN C PROGRAMMING

In C programming, a variable is a name that represents a memory location. It is used to store a value that can be changed during the execution of the program. Variables are essential for creating dynamic programs that can handle different inputs and outputs.

Scope of a Variable

The scope of a variable refers to the area of the program where the variable is accessible. In C programming, the scope of a variable can be determined by where it is declared.

Global Variables

A global variable is declared outside of any function and can be accessed by any function in the program. Global variables have a scope that extends throughout the entire program. Here's an example of a global variable declaration:

```
int globalVar = 10;
```

Local Variables

A local variable is declared inside a function and can only be accessed within that function. Local variables have a scope that is limited to the function where they are declared. Here's an example of a local variable declaration:

```
void myFunction() {
  int localVar = 5;
```

```
 // code here
}
```

Life of a Variable

The life of a variable refers to the duration of time that a variable exists in the program's memory. In C programming, the life of a variable can be determined by where it is declared.

Automatic Variables

An automatic variable is a local variable that is created when a function is called and destroyed when the function ends. Automatic variables have a life that is limited to the execution of the function. Here's an example of an automatic variable declaration:

```
void myFunction() {
 int autoVar = 5;
 // code here
}
```

Static Variables

A static variable is a local variable that retains its value between function calls. Static variables have a life that extends beyond the execution of the function. Here's an example of a static variable declaration:

```
void myFunction() {
 static int staticVar = 5;
 // code here
}
```

Example Program

Let's look at an example program that demonstrates the scope and life of variables in C programming:

```
#include <stdio.h>
int globalVar = 10;
void myFunction()
{
 int localVar = 5;
```

```
  static int staticVar = 5;
  printf("localVar = %d\n", localVar);
  printf("staticVar = %d\n", staticVar);
  localVar++;
  staticVar++;
}
int main() {
  int autoVar = 5;
  myFunction();
  printf("globalVar = %d\n", globalVar);
  printf("autoVar = %d\n", autoVar);
  myFunction();
  return 0;
}
```

In this program, we declare a global variable, a local variable, an automatic variable, and a static variable. We also define a function that modifies the values of the local and static variables. Finally, we call the function twice and print the values of all the variables.

When we run this program, we get the following output:

```
localVar = 5
staticVar = 5
globalVar = 10
autoVar = 5
localVar = 6
staticVar = 6
```

This output shows that the local and static variables are only accessible within the function, while the global and automatic variables are accessible throughout the program. It also shows that the automatic variable is destroyed and re-created every time the function is called, while the static variable retains its value between function calls.

10.6 MULTIFILE PROGRAMMING IN C PROGRAMMING

In larger C programs, it is often necessary to divide the code into multiple files for easier maintenance, better organization, and improved scalability. Multifile programming involves creating multiple

C source files that can be compiled and linked together to form a single executable file. In this chapter, we will discuss the basics of multifile programming in C programming and provide some examples to illustrate the concepts.

Creating Header Files

Header files contain declarations for functions, variables, and other types of data that are used across multiple source files. The declarations are placed in a header file with a .h extension, and the corresponding implementation is placed in a source file with a .c extension. The syntax for creating a header file is as follows:

```
#ifndef MYHEADER_H
#define MYHEADER_H
/* declarations go here */
#endif
```

The first two lines use the preprocessor directives #ifndef and #define to prevent the header file from being included multiple times in the same source file. The third line contains the declarations for functions, variables, and other types of data. The final line uses the #endif directive to close the #ifndef directive.

For example, consider the following header file named "myheader.h":

```
#ifndef MYHEADER_H
#define MYHEADER_H
int max(int a, int b);
#endif
```

This header file declares a function named "max" that takes two integer arguments and returns the maximum value.

Creating Source Files

Source files contain the implementation of functions, variables, and other types of data that are declared in the header files. The syntax for creating a source file is as follows:

```
#include "myheader.h"
/* implementation goes here */
int max(int a, int b) {
 return a > b ? a : b;
}
```

The first line includes the corresponding header file "myheader.h". The implementation of the "max" function is then provided in the source file.

Compiling and Linking Multiple Files

To compile and link multiple C source files, we use the gcc compiler with the following syntax:

```
gcc file1.c file2.c -o output
```

This command compiles the source files "file1.c" and "file2.c" and links them together to form an executable file named "output".

For example, consider the following source files "file1.c" and "file2.c":

```
// file1.c
#include "myheader.h"
int main() {
 int x = 10;
 int y = 20;
 int result = max(x, y);
 return 0;
}
// file2.c
#include "myheader.h"
/* implementation of max function */
```

The "main" function in "file1.c" uses the "max" function declared in the "myheader.h" header file. The implementation of the "max" function is provided in "file2.c".

To compile and link these source files together, we use the following command:

This command generates an executable file named "myprogram" that can be run from the command line.

Example: Multifile Programming in C

Sure, let's go through a detailed example of multifile programming in C to illustrate the concepts discussed in the previous chapter. The example consists of three source files: "main.c", "sum.c", and "sum.h".

The "sum.h" file contains the declarations for the "sum" function, which calculates the sum of two integers. The "sum.c" file contains the implementation of the "sum" function, while the "main.c" file contains the main function that calls the "sum" function and prints the result to the console.

Here's the code for "sum.h":

```
// sum.h
int sum(int a, int b);
```

The "sum" function takes two integer arguments and returns their sum.

Here's the code for "sum.c":

```
// sum.c
#include "sum.h"
int sum(int a, int b)
{
return a + b;
}
```

The implementation of the "sum" function is provided in "sum.c". The function simply adds the two integer arguments and returns the result.

Finally, here's the code for "main.c":

```
// main.c
#include <stdio.h>
#include "sum.h"
int main()
{
int x = 10;
int y = 20;
 int result = sum(x, y);
printf("The sum of %d and %d is %d\n", x, y, result);
return 0;
}
```

The "main" function in "main.c" includes the "stdio.h" header file for the "printf" function and the "sum.h" header file for the "sum" function. It then declares two integer variables "x" and "y" and initializes them to 10 and 20, respectively. The "sum" function is called with "x" and "y" as arguments, and the result is stored in the "result" variable. Finally, the "printf" function is called to print the result to the console.

To compile and link these source files together, we use the following command:

```
gcc main.c sum.c -o myprogram
```

This command generates an executable file named "myprogram" that can be run from the command line.

When the program is run, the output to the console should be:

The sum of 10 and 20 is 30

This example demonstrates how to use multifile programming in C to organize code into separate source files and create reusable functions that can be used across multiple source files. By dividing the code into multiple files, we can improve the maintainability and scalability of the program.

EXERCISE

Short Questions (2 marks each)

1. What is a command line argument?
2. How many file modes are available in Python?
3. What is the difference between read() and readline() methods in file handling?
4. What is the default file mode in Python's open() function?
5. How can you check if a file exists in Python?

Long Questions (5 marks each)

1. Explain the concept of command line arguments and how to use them in Python programs.
2. What are the different file modes available in Python? Explain each mode in detail with examples.
3. How can you open a file in Python and perform basic file operations like reading, writing, and appending data? Explain with examples.
4. What is file pointer in Python file handling? How can you move

the file pointer to a specific position in a file? Explain with examples.

5. How can you handle exceptions while working with files in Python? Explain with examples.

Multiple Choice Questions

1. What is the function used to open a file in Python?
   a) open()
   b) read()
   c) write()
   d) append()

Answer: a) open()

2. Which file mode is used to open a file for reading and writing?
   a) 'r'
   b) 'w'
   c) 'a'
   d) 'r+'

Answer: d) 'r+'

3. What is the difference between 'w' and 'a' file modes in Python?
   a) 'w' mode overwrites the existing file while 'a' mode appends data to the end of the file.
   b) 'w' mode creates a new file while 'a' mode appends data to the end of the existing file.
   c) 'w' mode appends data to the end of the file while 'a' mode overwrites the existing file.
   d) There is no difference between 'w' and 'a' file modes.

Answer: a) 'w' mode overwrites the existing file while 'a' mode appends data to the end of the file.

4. What is the purpose of file pointer in Python file handling?
   a) It indicates the current position in the file where the next read or write operation will take place.
   b) It is used to create a new file.
   c) It is used to delete a file.
   d) None of the above.

Answer: a) It indicates the current position in the file where the next read or write operation will take place.

5. Which method is used to read the entire contents of a file in Python?
   a) read()
   b) readline()
   c) readlines()
   d) None of the above.

Answer: a) read()

6. What happens when you try to open a file that does not exist in Python?
   a) It raises a FileNotFoundError exception.
   b) It creates a new file with the specified name.
   c) It returns None.
   d) It raises a SyntaxError exception.

Answer: a) It raises a FileNotFoundError exception.

7. Which file mode is used to open a file for writing only?
   a) 'r'
   b) 'w'
   c) 'a'
   d) 'x'

Answer: b) 'w'

8. Which method is used to write data to a file in Python?
   a) write()
   b) read()
   c) readline()
   d) None of the above.

Answer: a) write()

11

Basic Algorithms

11.1 INTRODUCTION

In computer science, algorithms are a set of instructions used to perform a specific task. They are used to solve problems and provide solutions to complex issues. This chapter will provide a comprehensive overview of basic algorithms and their applications in computer science.

Section 1: Sorting Algorithms

Sorting algorithms are a fundamental part of computer science, as they allow data to be ordered in a specific manner. Common sorting algorithms include:

- **Bubble sort**: a simple sorting algorithm that repeatedly steps through the list, compares adjacent elements and swaps them if they are in the wrong order.
- **Insertion sort**: an efficient sorting algorithm that builds the final sorted array one item at a time. It is much less efficient on large lists than more advanced algorithms such as quicksort, heapsort, or merge sort.
- **Quick sort**: a divide-and-conquer sorting algorithm that works by selecting a 'pivot' element from the array and partitioning the other elements into two sub-arrays, according to whether they are less than or greater than the pivot.

Section 2: Search Algorithms

Search algorithms are used to search for specific data within a larger data set. Common search algorithms include:

- **Linear search:** a simple search algorithm that checks each element of the list in sequence until a match is found or the end of the list is reached.
- **Binary search:** a more efficient search algorithm that works by dividing the search interval in half with each iteration, until the value is found or the search interval is empty.

Section 3: Graph Algorithms

Graph algorithms are used to analyse and manipulate networks and graphs. They have a wide range of applications, including social network analysis, transportation, and telecommunications. Common graph algorithms include:

- **Breadth-first search (BFS):** a graph traversal algorithm that visits all the vertices of a graph in breadth-first order, i.e., it explores all the vertices at distance 1 from the source vertex before visiting vertices at distance 2.
- **Depth-first search (DFS):** a graph traversal algorithm that visits all the vertices of a graph in depth-first order, i.e., it explores as far as possible along each branch before backtracking.

Section 4: Dynamic Programming Algorithms

Dynamic programming is a method for solving complex problems by breaking them down into smaller, overlapping sub-problems. It is used to optimize problems where the solution can be broken down into overlapping sub-problems and then recombined to form the solution. Common applications of dynamic programming include:

- **Fibonacci sequence:** a sequence of numbers where each number is the sum of the two preceding ones, starting from 0 and 1.
- **Longest common subsequence (LCS)**: a measure of the similarity between two sequences, defined as the longest sequence of elements that appear in both the original sequences in the same order.

In conclusion, this chapter has provided a basic overview of common algorithms used in computer science. Understanding the fundamentals of these algorithms is essential for developing efficient and effective solutions to real-world problems.

11.2 FIBONACCI SERIES

The Fibonacci series is a sequence of numbers in which each number is the sum of the two preceding ones. It is named after Leonardo of Pisa, who was known as Fibonacci, and was first introduced in his book Liber Abaci in 1202. The first two numbers in the series are usually 0 and 1, and the sequence goes 0, 1, 1, 2, 3, 5, 8, 13, 21, and so on.

Here is a pseudocode algorithm for generating the Fibonacci series:

Algorithm: Fibonacci

Input: A positive integer n

Output: The first n terms of the Fibonacci series

1. Initialize two variables, t1 and t2, to 0 and 1, respectively.
2. Print t1.
3. For i = 1 to n-1, do the following:
    - Calculate the next term as the sum of t1 and t2.
    - Set t1 to t2.
    - Set t2 to the next term.
    - Print t2.

Here is a program in C that implements the above algorithm to generate the Fibonacci series:

```
#include <stdio.h>
// Function to generate the Fibonacci series
void fibonacci(int n) {
 int i, t1 = 0, t2 = 1, nextTerm;
 printf("Fibonacci Series: ");
 for (i = 1; i <= n; ++i)
 {
 printf("%d, ", t1);
 nextTerm = t1 + t2;
 t1 = t2;
 t2 = nextTerm;
 }
}
```

```
int main() {
 int n;
 printf("Enter the number of terms: ");
 scanf("%d", &n);
 fibonacci(n);
 return 0;
}
```

Output:

```
Enter the number of terms: 10
Fibonacci Series: 0, 1, 1, 2, 3, 5, 8, 13, 21, 34,
```

11.3 LINEAR SEARCH

Linear search, also known as sequential search, is a simple searching algorithm that iterates through a collection of elements, one by one, and returns the first element that matches the search key. It is the simplest form of search and is often used when the data is not sorted and the size of the collection is small.

Here is a pseudocode algorithm for linear search:

Algorithm: Linear Search

Input: An array A of n elements and a search key k

Output: The index of the first element in A that is equal to k, or -1 if k is not found

1. For i = 0 to n-1, do the following:
   a. If A[i] is equal to k, return i
2. Return -1

Here is a program in C that implements the above algorithm for linear search:

```
#include <stdio.h>
// Function to perform linear search
int linearSearch(int array[], int n, int key) {
 int i;
 for (i = 0; i < n; i++) {
 if (array[i] == key)
 return i;
```

```
  }
  return -1;
 }
 int main() {
  int array[100], n, key, result;
  printf("Enter the number of elements: ");
  scanf("%d", &n);
  printf("Enter the elements: ");
  for (int i = 0; i < n; i++) {
  scanf("%d", &array[i]);
  }
  printf("Enter the search key: ");
  scanf("%d", &key);
  result = linearSearch(array, n, key);
  if (result == -1)
  printf("%d is not found in the array.\n", key);
  else
  printf("%d is found at index %d.\n", key, result);
  return 0;
 }
```

Output:

```
Enter the number of elements: 5
Enter the elements: 1 2 3 4 5
Enter the search key: 3
3 is found at index 2.
```

11.4 BINARY SEARCH

Binary search is a fast-searching algorithm that works on sorted arrays. Unlike linear search, which checks each element in the array one by one, binary search takes advantage of the sorted order and reduces the number of elements to check by half after each comparison. It works by repeatedly dividing the search interval in half until the target value is found or it is clear that the value is not present in the array.

Here is a pseudocode algorithm for binary search:

Algorithm: Binary Search

Input: An array A of n sorted elements and a search key k

Output: The index of the first element in A that is equal to k, or -1 if k is not found

1. Set low = 0 and high = n-1
2. While low <= high, do the following:
   a. Calculate mid = (low + high) / 2
   b. If A[mid] is equal to k, return mid
   c. If A[mid] is less than k, set low = mid + 1
   d. If A[mid] is greater than k, set high = mid - 1
3. Return -1

Here is a program in C that implements the above algorithm for binary search:

```
#include <stdio.h>
// Function to perform binary search
int binarySearch(int array[], int n, int key) {
 int low = 0, high = n - 1, mid;
 while (low <= high) {
 mid = (low + high) / 2;
 if (array[mid] == key)
 return mid;
 else if (array[mid] < key)
 low = mid + 1;
 else
 high = mid - 1;
 }
 return -1;
}
int main() {
 int array[100], n, key, result;
 printf("Enter the number of elements: ");
 scanf("%d", &n);
```

```
    printf("Enter the elements in ascending order: ");
    for (int i = 0; i < n; i++) {
    scanf("%d", &array[i]);
    }
    printf("Enter the search key: ");
    scanf("%d", &key);
    result = binarySearch(array, n, key);
    if (result == -1)
    printf("%d is not found in the array.\n", key);
    else
    printf("%d is found at index %d.\n", key, result);
    return 0;
}
```

Output:

```
Enter the number of elements: 5
Enter the elements in ascending order: 1 2 3 4 5
Enter the search key: 3
3 is found at index 2.
```

11.5 SORTING TECHNIQUES

Sorting algorithms are a set of algorithms that are used to arrange elements in a specific order, either in ascending or descending order. Sorting algorithms are a fundamental concept in computer science and are widely used in various applications, such as data analysis, database management, and computer graphics. There are many different sorting algorithms, each with its own advantages and disadvantages in terms of time complexity, space complexity, and stability.

Bubble Sort

One of the simplest sorting algorithms is Bubble Sort. Bubble sort is a comparison-based sorting algorithm that repeatedly steps through the list, compares adjacent elements and swaps them if they are in the wrong order. The algorithm gets its name from the way smaller elements "bubble" to the top of the list.

Here is a pseudocode algorithm for Bubble Sort:

Algorithm: Bubble Sort

Input: An array A of n elements

Output: A sorted in ascending order

1. For i = 0 to n-2, do the following:
   a. For j = 0 to n-2-i, do the following:
      i. If A[j] > A[j+1], swap A[j] and A[j+1]
2. Return A

Here is a program in C that implements the above algorithm for Bubble Sort:

```
#include <stdio.h>
// Function to perform bubble sort
void bubbleSort(int array[], int n) {
 int temp;
 for (int i = 0; i < n - 1; i++) {
 for (int j = 0; j < n - i - 1; j++) {
 if (array[j] > array[j + 1]) {
 temp = array[j];
 array[j] = array[j + 1];
 array[j + 1] = temp;
 }
 }
 }
}

 int main() {
  int array[100], n;
  printf("Enter the number of elements: ");
  scanf("%d", &n);
  printf("Enter the elements: ");
  for (int i = 0; i < n; i++) {
  scanf("%d", &array[i]);
  }
```

```
    bubbleSort(array, n);
    printf("Sorted array: ");
    for (int i = 0; i < n; i++) {
    printf("%d ", array[i]);
    }
    printf("\n");
    return 0;
}
```

Output:

```
Enter the number of elements: 5
Enter the elements: 5 4 3 2 1
Sorted array: 1 2 3 4 5
```

Insertion Sort

Insertion sort is a simple sorting algorithm that builds the final sorted array (or list) one item at a time. It is much less efficient on large lists than more advanced algorithms such as quicksort, heapsort, or merge sort. However, insertion sort provides several advantages:

- **Simple implementation:** Insertion sort is easy to understand and simple to implement.
- **Efficient for small data sets:** Insertion sort is efficient for small data sets, with a time complexity of O(n) when the data is already sorted.
- **Stable sorting:** Insertion sort is a stable sort, meaning that it maintains the relative order of equal elements.
- **Adaptive**: Insertion sort is an adaptive sort, meaning that it performs better for nearly sorted or partially sorted arrays.

Here is a pseudocode algorithm for Insertion Sort:

Algorithm: Insertion Sort

Input: An array A of n elements

Output: A sorted in ascending order

1. For i = 1 to n-1, do the following:
    a. key = A[i]
    b. j = i-1

c. While j >= 0 and A[j] > key, do the following:

i. A[j + 1] = A[j]

ii. j = j - 1

d. A[j + 1] = key

2. Return A

Here is a program in C that implements the above algorithm for Insertion Sort:

```
#include <stdio.h>
// Function to perform insertion sort
void insertionSort(int array[], int n) {
 int key, j;
 for (int i = 1; i < n; i++) {
 key = array[i];
 j = i - 1;
 while (j >= 0 && array[j] > key) {
 array[j + 1] = array[j];
 j = j - 1;
 }
 array[j + 1] = key;
 }
}
int main() {
 int array[100], n;
 printf("Enter the number of elements: ");
 scanf("%d", &n);
 printf("Enter the elements: ");
 for (int i = 0; i < n; i++) {
 scanf("%d", &array[i]);
 }
 insertionSort(array, n);
 printf("Sorted array: ");
 for (int i = 0; i < n; i++) {
```

```
    printf("%d ", array[i]);
  }
  printf("\n");
  return 0;
}
```

Output:

```
Enter the number of elements: 5
Enter the elements: 5 4 3 2 1
Sorted array: 1 2 3 4 5
```

Selection Sort

Selection sort is a simple sorting algorithm that works by repeatedly finding the minimum element from the unsorted part of the array and putting it at the beginning. The algorithm divides the input into two parts: the sorted part at the left end and the unsorted part at the right end. Initially, the sorted part is empty and the unsorted part is the entire array.

Here is a pseudocode algorithm for Selection Sort:

Algorithm: Selection Sort

Input: An array A of n elements

Output: A sorted in ascending order

1. For i = 0 to n-2, do the following:
    a. Find the minimum element in the unsorted part of the array
    b. Swap the found minimum element with the first element in the unsorted part
2. Return A

Here is a program in C that implements the above algorithm for Selection Sort:

```
#include <stdio.h>
// Function to perform selection sort
void selection Sort(int array[], int n) {
  int minIndex;
  for (int i = 0; i < n-1; i++) {
  minIndex = i;
```

```
for (int j = i + 1; j < n; j++) {
if (array[j] < array[minIndex]) {
minIndex = j;
}
}
int temp = array[minIndex];
array[minIndex] = array[i];
array[i] = temp;
}
}
int main() {
int array[100], n;
printf("Enter the number of elements: ");
scanf("%d", &n);
printf("Enter the elements: ");
for (int i = 0; i < n; i++) {
scanf("%d", &array[i]);
}
selectionSort(array, n);
printf("Sorted array: ");
for (int i = 0; i < n; i++) {
printf("%d ", array[i]);
}
printf("\n");
return 0;
}
```

Output:

```
Enter the number of elements: 5
Enter the elements: 5 4 3 2 1
Sorted array: 1 2 3 4 5
```

11.6 SQUARE ROOT OF A NUMBER

The square root of a number is a value that, when multiplied by itself, equals the original number. For example, the square root of 9 is 3, since 3 x 3 = 9.

Here is a pseudocode algorithm for finding the square root of a number:

Algorithm: Square root of a number

Input: A positive number x

Output: The square root of x

1. Set a starting value for the square root, e.g., x/2
2. Repeat the following steps until the desired precision is achieved:

   a. Compute a new estimate using the formula:

   new Estimate = (previous Estimate + x/previous Estimate) / 2

   b. Update previous Estimate with new Estimate
3. Return previous Estimate

Here is a program in C that implements the above algorithm for finding the square root of a number:

```
#include <stdio.h>
#include <math.h>
// Function to find the square root of a number
double squareRoot(double x) {
 double estimate = x / 2;
 double newEstimate;
 do {
 newEstimate = (estimate + x/estimate) / 2;
 estimate = newEstimate;
 } while (fabs(estimate - newEstimate) > 0.00001);
 return estimate;
}
int main() {
 double x;
```

```
printf("Enter a positive number: ");
scanf("%lf", &x);
printf("Square root of %.2lf is %.2lf\n", x, squareRoot(x));
return 0;
}
```

Output:

```
Enter a positive number: 9
Square root of 9.00 is 3.00
```

11.7 ARRAY ORDER REVERSAL

Array order reversal refers to flipping the order of elements in an array. For example, if an array A has the elements [1, 2, 3, 4], the reversed array will be [4, 3, 2, 1].

Here is a pseudocode algorithm for array order reversal:

Algorithm: Array Order Reversal

Input: An array A of size n

Output: Reversed array A

1. Set i = 0 and j = n-1
2. Repeat the following steps while i < j:
   a. Swap A[i] and A[j]
   b. Increment i by 1 and decrement j by 1
3. Return A

Here is a program in C that implements the above algorithm for array order reversal:

```
#include <stdio.h>
// Function to reverse an array
void reverseArray(int A[], int n) {
 int i = 0;
 int j = n-1;
 int temp;
 while (i < j) {
 temp = A[i];
 A[i] = A[j];
```

```
 A[j] = temp;
 i++;
 j--;
 }
}
int main() {
 int n, i;
 printf("Enter the size of the array: ");
 scanf("%d", &n);
 int A[n];
 printf("Enter %d elements: ", n);
 for (i = 0; i < n; i++) {
 scanf("%d", &A[i]);
 }
 reverseArray(A, n);
 printf("Reversed array: ");
 for (i = 0; i < n; i++) {
 printf("%d ", A[i]);
 }
 printf("\n");
 return 0;
}
```

Output:

```
Enter the size of the array: 5
Enter 5 elements: 1 2 3 4 5
Reversed array: 5 4 3 2 1
```

11.8 STRING REVERSAL

String reversal refers to flipping the order of characters in a string. For example, if a string s has the value "hello", the reversed string will be "olleh".

Here is a pseudocode algorithm for string reversal without using any inbuilt functions:

Algorithm: String Reversal (without inbuilt function)

Input: A string s of length n

Output: Reversed string s

1. Set i = 0 and j = n-1
2. Repeat the following steps while i < j:
   a. Swap s[i] and s[j]
   b. Increment i by 1 and decrement j by 1
3. Return s

Here is a program in C that implements the above algorithm for string reversal without using any inbuilt functions:

```
#include <stdio.h>
#include <string.h>
// Function to reverse a string
void reverseString(char s[]) {
 int i = 0;
 int j = strlen(s) - 1;
 char temp;
 while (i < j) {
 temp = s[i];
 s[i] = s[j];
 s[j] = temp;
 i++;
 j--;
 }
}

int main() {
 char s[100];
 printf("Enter a string: ");
 scanf("%s", s);
 reverseString(s);
 printf("Reversed string: %s\n", s);
 return 0;
}
```

Output:

```
Enter a string: hello
Reversed string: olleh
```

And here is a pseudocode algorithm for string reversal using an inbuilt function strrev():

Algorithm: String Reversal (with inbuilt function)

Input: A string s of length n

Output: Reversed string s

1. Call the strrev(s) function from the string.h library
2. Return s

Here is a program in C that implements the above algorithm for string reversal using an inbuilt function:

```
#include <stdio.h>
#include <string.h>
int main() {
 char s[100];
 printf("Enter a string: ");
 scanf("%s", s);
 strrev(s);
 printf("Reversed string: %s\n", s);
 return 0;
}
```

Output:

```
Enter a string: hello
Reversed string: olleh
```

EXERCISE

Short Questions (2 marks each)

1. What is the time complexity of finding factorial using recursion?
2. What is the nth number in the Fibonacci series?

3. What is linear search?
4. What is the worst-case time complexity of Bubble sort?
5. What is the time complexity of finding the square root of a number using binary search?

Long Questions (5 marks each)

1. Explain how to find the nth number in the Fibonacci series using recursion.
2. Describe the Bubble sort algorithm and explain its time complexity.
3. Explain how to reverse the order of elements in an array.
4. Discuss the Insertion sort algorithm and compare its time complexity with Bubble sort.
5. Explain how to find the square root of a number using the binary search algorithm.

Multiple Choice Questions

1. Which of the following algorithms has the worst-case time complexity of O(n^2)?
   a) Bubble sort
   b) Insertion sort
   c) Selection sort
   d) Quick sort

Answer: a) Bubble sort

2. Which of the following algorithms is used to find the square root of a number?
   a) Linear search
   b) Binary search
   c) Bubble sort
   d) Insertion sort

Answer: b) Binary search

3. Which of the following algorithms has the best-case time complexity of O(n)?
   a) Bubble sort
   b) Insertion sort

c) Selection sort
d) Quick sort

Answer: b) Insertion sort

4. Which of the following is the correct order of steps in the Bubble sort algorithm?
   a) Compare adjacent elements, swap if necessary, repeat for all elements
   b) Find the minimum element, swap with the first element, repeat for all elements
   c) Find the maximum element, swap with the last element, repeat for all elements
   d) None of the above

Answer: a) Compare adjacent elements, swap if necessary, repeat for all elements

5. Which of the following algorithms is used to reverse the order of elements in an array?
   a) Bubble sort
   b) Insertion sort
   c) Selection sort
   d) None of the above

Answer: d) None of the above (Array reversal can be done by iterating through the array and swapping the first and last elements, second and second-to-last elements, and so on.)

6. Which of the following algorithms is used to find a specific element in an unsorted array?
   a) Linear search
   b) Binary search
   c) Bubble sort
   d) Insertion sort

Answer: a) Linear search

7. Which of the following algorithms has the worst-case time complexity of O(n log n)?
   a) Bubble sort
   b) Insertion sort

c) Selection sort

d) Merge sort

Answer: d) Merge sort

8. Which of the following algorithms is used to reverse the order of characters in a string?

   a) Bubble sort

   b) Insertion sort

   c) Selection sort

   d) None of the above

Answer: d) None of the above (String reversal can be done by iterating through the string and appending the characters in reverse order to a new string.)

9. Which of the following algorithms has the best-case time complexity of O(n^2)?

   a) Bubble sort

   b) Insertion sort

   c) Selection sort

   d) Quick sort

Answer: a) Bubble sort

10. Which of the following algorithms is used to find the nth number in the Fibonacci series?

   a) Linear search

   b) Binary search

   c) Recursion

   d) Bubble sort

Answer: c) Recursion

Appendix: C99 Extensions

"An In-Depth Look at C Standard Libraries"

C is a widely-used programming language known for its low-level control and efficiency. One of the benefits of using C is access to its standard libraries, a collection of functions and macros that provide common tasks for C programmers. In this chapter, we will delve into the standard C libraries and their functions, exploring the most essential and commonly used tools for C programmers in depth.

1. **Overview of the Standard Libraries:** The C standard libraries are a collection of functions and macros that provide common programming tasks such as memory management, string manipulation, input/output operations, and mathematics. These libraries are specified by the C standard and are available on all platforms that support the C language. The libraries are grouped into headers, each containing related functions. Some of the most commonly used headers include:
   - stdio.h: for input and output operations (e.g. printf, scanf)
   - string.h: for string manipulation functions (e.g. strcpy, strcmp)
   - math.h: for mathematical operations (e.g. sin, cos)
   - stdlib.h: for general utility functions (e.g. malloc, exit)
   - time.h: for time-related operations (e.g. clock, time)
2. **Memory Management:** Memory management is an important aspect of C programming. Functions like malloc, calloc, and realloc allow dynamic memory allocation, enabling the program to obtain memory space as required. The function free releases previously allocated memory. Additionally, the standard library provides functions for determining the amount of available memory, such as mallinfo and malloc_stats.

3. **String Manipulation:** String manipulation is a common task in C programming. The string.h header provides functions for tasks like copying, concatenating, and comparing strings. For example, strcpy copies one string to another, strcat concatenates two strings, and strcmp compares two strings. Additionally, there are functions for searching and transforming strings, such as strstr, strtok, and strrev.
4. **Input and Output:** The stdio.h header provides functions for input and output operations. The most commonly used functions include printf for printing to the screen and scanf for reading input from the user. Additionally, there are functions for reading and writing to and from files, such as fopen, fclose, fprintf, and fscanf. The standard library also provides functions for controlling and redirecting input and output streams, such as setbuf and freopen.
5. **Mathematics:** The math.h header provides a range of mathematical functions, including trigonometric functions (sin, cos, tan), exponential functions (exp, log), and power functions (pow). These functions are useful for a variety of applications, including scientific and engineering calculations. Additionally, the standard library provides functions for performing complex arithmetic, such as cmath and complex.h.
6. **Time and Date:** The time.h header provides functions for working with time and date. The most commonly used functions include time, which returns the current time, and clock, which returns the number of clock ticks since the program started. Additionally, there are functions for working with the calendar, such as localtime and gmtime, which convert time values into human-readable date and time representations.

Made in the USA
Monee, IL
13 September 2026

62980093R00108